OTHER TITLES OF INTEREST

ISO 9000: An Implementation Guide for Small and Mid-Sized Businesses

The 90-Day ISO 9000 Manual: Basics Manual and Implementation Guide

QS-9000 Handbook

Principles of Total Quality

Quality Improvement Handbook: Team Guide to Tools and Techniques

Total Quality Management: Text, Cases, and Readings, 2nd Edition

Introduction to Modern Statistical Quality Control and Management

Focused Quality: Managing for Results

The Executive Guide to Implementing Quality Systems

Total Quality Service: Principles, Practices, and Implementation

Total Quality in Purchasing and Supplier Management

INSIDE ISO 14000

The Competitive Advantage of Environmental Management

DON SAYRE

StL
St. Lucie Press
Delray Beach, Florida

Printed and bound in the U.S.A. Printed on acid-free paper.
10 9 8 7 6 5 4 3 2 1

ISBN 1-57444-028-4

Direct all inquiries to St. Lucie Press, Inc., 100 E. Linton Blvd., Suite 403B, Delray Beach, Florida 33483.

Phone: (407) 274-9906
Fax: (407) 274-9927

S^{t}L

Published by
St. Lucie Press
100 E. Linton Blvd., Suite 403B
Delray Beach, FL 33483

Table of Contents

PREFACE

St. Lucie Press approached me with the idea of a primer on ISO 14000, the upcoming environmental management systems standard soon to be finalized by the International Organization for Standardization. I knew the committee draft of the standard and was (still am) in love with the concepts and promises it makes. So, I said, sure, let's do it. This primer is the result of our initial conversation.

This primer is written to present the principles discussed in the ISO 14000 series of standards and guidelines without duplicating exact text. I firmly believe that the content of the final series will be close to what has been drafted, balloted and agreed to so far. The daughter standards are changing even as I write, but the mother and family of ISO 14000 will change very little.

The function of this primer is to promote adoption of the specification and guidance documents within the ISO 14000 series. I act as the messenger, attempting to bring the word down from the mountain top so you can see it, hold it, consider its wisdom, and act in good conscience to protect and preserve the environment. I try to tell you everything I know of value on establishing, implementing, maintaining, and evaluating an effective environmental management system.

For the price of this book, you get everything you need to know. The key to good environmental management is the systems approach, summarized as follows:

- commitment, planning and policy
- leadership, communications and resources
- controls, procedures and training
- audits, reviews and improvements
- stakeholder participation and motivation

As the subtitle implies, the ISO 14000 methodology for environmental management is a competitive approach. It provides the necessary requirements and recommendations for any organization to develop and implement a cost-effective system of management. And above all that, it can ensure sustainable development for everyone in the world if practiced with rigor and some discipline. Due diligence is an obligation, but it's also an opportunity and can be to your advantage as well as to everyone else's.

Management is an art, a science. I believe you can use the ISO 14000 series in a number of profitable and prospering ways:

- become ISO 14000 certified
- satisfy your customers
- meet the increasing requirements of society
- beat your competition

With an effective management system, you have the competitive advantage and can see your enemies before they attack. Once your system's in place, you'll know for sure what you can do *to* and *for* the environment. Then you'll have the power to make decisions and act.

The world is still in the Industrial Revolution—always has been, always will be. But now, with the shrinking resources, the "smaller planet," the revolution needs to include two more arsenals—one for the protection and preservation of the environment and the other for the sustainable development of life and improving the ecosystem.

What you're about to read is a primer on the way to stock those arsenals and deploy them for the most advantage. The International Organization for Standardization (ISO) has a master scheme in mind. They call it ISO 14000, the standard and guidelines for effective environmental management systems.

This primer is your start. It introduces readers to the concepts, the tools, and the ways to break through and mentions some of the potential potholes and pockets of failure to steer clear of.

You are the captain of your ship. Regulations and standards like ISO 14000 are simply maps and compasses to follow. When you adopt one, you go that route. And when you can't, you say so and keep on going. Nothing in life mandates how you "do" life. Standards like ISO 14000 are commitments you make. To create and adopt an environmental management standard that fits its specifications more or less and its guidance is your option.

This primer follows the format of the latest information available on ISO 14000. The chapters and sub-elements are numbered to correspond to and provide guidance to the draft standard, with additional guidance from personal observation.

I asked two colleagues with excellent credentials to take a close look at the primer and share their opinions and ideas. First, the wisdom of Gary Corrigan, currently a controversial but respected Lead Auditor with the new Lockheed Martin Utility Services, Inc. Gary is responsible for assuring the quality of operations and performance at the U.S. Enrichment Corporation's gaseous diffusion plant at Paducah, Kentucky. Gary has a Masters in Environmental Science from Washington State University and has worked the nuclear power plant construction world for years.

> For the past 15 years, I have been performing quality assurance activities at both commercial nuclear and U.S. Department of Energy facilities. Over the period, I've

> noticed two problems that consistently come to the surface.... First is the generalization that *all* environmental problems are the same. Hence, one environmental program is generated, then 'force-fit' to the remediation activities for *all* sites, regardless of their individual needs.
>
> This primer addresses the issue head on. It leaves *environmental* systems to the side and concentrates on the overall *management* system. Here, a complete system package is promulgated to be applied for any environmental activity.
>
> The second problem is organizations not knowing or understanding the expectations of their regulator at the beginning of the process. Knowing what your regulators expect from the start promotes a sense of team-playing and always results in better managed projects. Knowing your regulators, their expectations, the laws of the land—all are dealt with in this primer. It is a far more than an informative guide to understanding and implementing environmental management as a competitive advantage.

George Sherwood is in the hierarchy of the U.S. Department of Energy. He reports to the Secretary of Energy, with only one level between him and Hazel O'Leary.

> In the last five years, I have been on the Staff of the Secretary's Advisory Committee on Nuclear Facility Safety, in the Office of Nuclear Energy Self Assessment and the Office of International Nuclear Safety.
>
> This primer should be very useful to those busy executives who have recently gained environmental management responsibilities. This primer will really help them to understand what they need to know and do. In addition to what is included in the primer, I have two additional recommendations.

> First, make arrangements for occasional independent external review of your environmental management program. And second, get to know your responsible regulatory personnel and *their* expectations of your environmental management system. As Gary mentioned, the lack of understanding your regulator is a serious problem that continually resurfaces.
>
> The benefits of both of these recommendations should be obvious. Outside management review always sheds light on better management practices and knowing the individuals who govern your organization's very existence empowers you to manage wisely.

Your environmental management needs to be systematic and personal. You need to accept responsibility as an organization for environmental impact. You need to fine tune your senses to stay alert to any problems you may cause the environment or the sustainable development of mankind. You need to feel your way around in your industry and your regulatory responsibilities and obligations. Stay in touch with reality. Know where you are at all times, where you want to go, where you came from.

Now, on into ISO 14000 to prime your environmental management pump for future operation. Trust me, this ISO standard will move us forward from the past into a better environment for all.

Manage the system or it's chance and chaos.

ACKNOWLEDGMENTS

Writing this primer on the future of environmental management depended on reading everything available and talking with as many friends and colleagues as possible. There are a few who merit mention. My appreciation is sincere.

Deep thanks to St. Lucie Press for their vision, to the Paducah, Kentucky library for maintaining a very up-to-date collection, and to JR's Executive Inn in Paducah for the support and friendship of the entire staff (especially to Randy Armstrong, Gerald, and to Leelee, who kept the room clean without making me leave the keyboard).

Warm thanks to George Sherwood of the U.S. Department of Energy and Gary Corrigan of Lockheed Martin Utility Services, Inc. for their support and encouragement. And thanks to Pete Hunter with MAC Technical Services Company in Germantown, Maryland for first turning me on to ISO 14000 and keeping the reference material current.

Devoted thanks to family and friends who gave me much inspiration, especially Sarah, my divine but devilish daughter in Florence, South Carolina, who gave me the encouragement to complete the manuscript as quickly as I could and cover everything well.

And strong kudos to my colleagues in the ASQC Energy and Environmental Division for making our Division the spokesperson on ISO 14000 for ASQC.

About the Author

Don Sayre has been involved in environmental, safety, health, and quality management since 1970. A recognized expert on ISO 14000, as well as ISO 9000, and a trained ISO 9000 Lead Auditor, Don is often asked to review new ISO products.

Don has provided management support to the U.S. Department of Energy for almost half of his career. He served as Field Engineer on the Fast Flux Test Facility Liquid Metal Breeder Reactor during its construction. He has performed audits, reviews, and surveillance on the Hanford Atomic Energy Reservation and analyzed environmental and operational occurrences at the Rocky Flats Plant for root causes.

Don has worked in construction and operation of nuclear power plants, one independent spent fuel storage installation, fossil power plants, and in the chemical and plastics industry. He is a Senior Member and Regional Councilor with the ASQC Energy and Environmental Division (as well as writer, producer and narrator of the first ASQC video, *Energy and Environmental Stewardship*, available through ASQC and the Division).

Don is an ASQC Certified Quality Auditor and ASME NQA-1 Lead Auditor, the benchmark standard for the U.S. nuclear industry. He works as a consultant offering management support an promoting best management practices across the United States and will soon do the same in the Ukraine and Russia with the U.S. Department of energy on emergency response preparedness.

To all those compelled to follow the best practices for environmental management with a particular interest in satisfying the requirements and expectations of the new environmental management standard, ISO 14000.

INTRODUCTION

ISO 14000—Environmental Management Systems: General Guidelines on Principles, Systems and Supporting Techniques

0 The Environmental Management Principles and Elements of ISO 14000

What now? Yet another international standard?

Wait! This one will actually benefit everyone. It advocates:

- **sustainable development for each and every nation and**
- **sustainable development for each and every person.**

It promotes the principles and practices essential to the competitive advantage of sound environmental performance:

- **resource allocation,**
- **responsibility and accountability, and**
- **continuous performance evaluation for improvement.**

So far, nothing new. Its predecessor, the ISO 9000 series on quality management systems, set the precedent:

We must pay attention to the basics of business if we want to survive and continue to compete.

Now, the International Organization for Standardization has packaged the rules of the game for effective environmental management.[1]

The International Organization for Standardization (ISO) has well over 100 member countries. Headquartered in Geneva, Switzerland, its purpose is to develop and promote standards worldwide to assure consistency in management systems. The work of ISO is done through hundreds of Technical Committees (TCs), including the American National Standards Institute (ANSI), the U.S. representative to ISO. The U.S. representative to ANSI is ASQC, formerly the American Society for Quality Control.

Inside the ISO 14000 series of environmental management standards, the basics of how to establish, document, implement, and maintain an effective environmental management system are defined. The series is based on a number of national standards from countries in Europe and the United States.

In the near future, compliance with ISO 14000 will be mandated by customers and even now, compliance offers a competitive advantage to suppliers. A commitment to ISO 14000 is a definite indicator that an organization is dedicated to the principles and elements of the standard, with an eye on sustainable development and environmental performance. In other words, the organization is making sure activities, products, and services are good for mankind and the world.

This new standard is derived somewhat from British Standard 7750, the specification for environmental management systems recognized worldwide as a foundation for sound environmental performance.[2] Comparisons to both the ISO 9000 series and BS 7750 are noted throughout this book where they increase the reader's understanding of the principles and tools for environmental management and environmental performance. In addition, the 1994 American National Standard, ANSI/ASQC E4, is cited where its contents might enhance an understanding the elements of ISO 14000.[3]

This book explores each of the management system principles and elements decreed and promoted by the ISO 14000 series. It describes the management system elements and specifications associated with each, plus the recommended guidelines.

The Table of Contents of ISO 14000 are shown below.[4] It is doubtful the standard will change much following the review of the Draft International Standard.[5] It is understood, however, that the standard's identification number may change from "14000" to a new number to avoid confusion.[6]

Also included in this book are notes on the February 1995 meeting of the Working Group on ISO 14001 and the resulting guidance on use of the specification. Relationships to other standards, such as ISO 9000, BS 7750, and ANSI/ASQC E4 are given where pertinent. In an attempt to make this primer as useful as possible, only information I am absolutely sure of has been included—nothing speculative.

0 INTRODUCTION
  0.1 Overview
  0.2 Benefits of Having an Environmental Management System
1. SCOPE
2. NORMATIVE REFERENCES
3. DEFINITIONS
4. ENVIRONMENTAL MANAGEMENT SYSTEM (EMS) PRINCIPLES AND ELEMENTS
  4.1 HOW TO START: COMMITMENT AND POLICY
    4.1.1 General
    4.1.2 Top Management Commitment and Leadership
    4.1.3 Initial Environmental Review
    4.1.4 Environmental Policy
  4.2 PLANNING
    4.2.1 General
    4.2.2 Identification of Environmental Aspects and Evaluation of

Associated Environmental Impacts
4.2.3 Legal Requirements
4.2.4 Internal Performance Criteria
4.2.5 Environmental Objectives and Targets
4.2.6 Environmental Management Plans and Program
4.3 IMPLEMENTATION
4.3.1 General
4.3.2 Ensuring Capability
4.3.2.1 Resources - Human, Physical, and Financial
4.3.2.2 EMS Alignment and Integration
4.3.2.3 Accountability and Responsibility
4.3.2.4 Environmental Values and Motivation
4.3.2.5 Knowledge, Skills and Training
4.3.3 Support Action
4.3.3.1 Communication and Reporting
4.3.3.2 EMS Documentation
4.3.3.3 EMS Records and Information Management
4.3.3.4 Operational Controls
4.3.3.5 Emergency Preparedness and Response
4.4 MEASUREMENT AND EVALUATION
4.4.1 General
4.4.2 Measuring and Monitoring (Ongoing Performance)
4.4.3 Audits of Environmental Management Systems
4.5 REVIEW AND IMPROVEMENT
4.5.1 General
4.5.2 Review of the Environmental Management System
4.5.3 Corrective and Preventive Action
4.5.4 Continual Improvement
Appendix A EXAMPLES OF INTERNATIONAL ENVIRONMENTAL GUIDING PRINCIPLES
A1 The Rio Declaration On Environment and Development
A2 International Chamber of Commerce (ICC) Business Charter for Sustainable Development

The format of this book closely follows that of the ISO 14000 series to keep things simple. However, information from ISO 14001 and other standards in the series has been incorporated into the discussion of ISO 14000 elements when appropriate. The same goes for elements of the 1994 ANSI/ASQC E4 Standard on environmental management, BS 7750, and other ISO standards.

Both ISO 14000 and its predecessor, BS 7750, evolved from increasing concerns over the state of the environment now and tomorrow. Look at how ISO 14000 begins:

ISO 14000

As concern grows for maintaining and improving the quality of the environment and protecting human health, organisations of all sizes are increasingly turning their attention to the potential impacts of their activities, products and services.

The environmental performance of an organisation is of increasing importance to internal and external interested parties.

Achieving sound environmental performance requires organisational commitment to a systematic approach and to continual improvement.

Now, take a look at the beginning of BS 7750. It is too similar to ignore the relationship of the international family of environmental management standards. Both share the same expectations, the same credo, the same philosophical and practical approach to effective management for the sake of sustainable development. Both have in mind compliance with environmental regulations and shareholder requirements and desires.

Challenge is "the consistent stimulation of excitement, enthusiasm, and dedication" (Richardson and Thayer 1993). Environmental management is just that in today's reality and in the pursuit of tomorrow's. These ISO 14000 standards offer all of the components of challenge.

BS 7750

Organisations of all kinds are increasingly concerned to achieve and demonstrate sound environmental performance. They do so in the context of increasingly stringent legislation, the development of economic and other measures to foster environmental protection, and a general growth of concern about environmental affairs.

British Standard 7750 was prepared under the direction of the Environmental Management Standards Policy Committee. It is a specification for an environmental management system. The standard is designed to ensure and demonstrate compliance with stated environmental policies and objectives. It also provides guidance on implementation of the specification within an overall management system.

The standard claims to be for any organization. Directions are given on establishing an effective management system for sound environmental performance and environmental auditing. Some 500 participants, including 230 implementing organizations, contributed to the pilot study of the standard's feasibility.

British Standard 7750, like ISO 14000, establishes no absolute requirements for environmental performance. It does, however, require compliance with applicable legislation and regulations, with a commitment to continual improvement. That means there is no one set way to carry out activities. Your interpretation and approach to environmental performance may be different from others in your industry. And rightfully so.

Environmental management system specification ISO 14001 begins very much like both ISO 14000 and BS 7750 with the same concepts of growing environmental concerns, the need to demonstrate sound environmental performance, and the need for tighter controls of environmental impacts. Remember, as in the relationship of ISO 9000 and ISO 9001, the first being voluntary and the second being a contractual obligation, ISO 14001 is like ISO 9001. Future contracts will specify whether an ISO 14001 environmental management system is required.

If a company or organization starts on the transition to an ISO environmental management system by following ISO 14000, the switch to an ISO 14001 system will be easier, less demanding on resources, and thus a logical progression. Here's how ISO 14001 introduces itself:

ISO 14001

Organisations of all kinds are increasingly concerned to achieve and demonstrate sound environmental performance by controlling the impact of their activities, products, and services on the environment, taking into account their environmental policy and objectives. They do so in the context of increasingly stringent legislation, the development of economic policies and other measures to foster environmental protection, and a general growth of concern from interested parties about environmental matters including sustainable development.

The ANSI/ASQC E4 Standard begins a little differently, having less of a scope and purpose but as much of a potential contribution to environmental improvement. The E4 standard is of modular design. It describes the quality management elements generally common to environmental programs, regardless of their scope, in one module. Other modules contain the quality system elements applicable to specific technical areas. It's up to the user to determine specific applicability to the collection and evaluation of environmental data or to the design, construction, and operation of environmental technology.

Note that the E4 standard is fairly equivalent to one of the strictest consensus standards ever written and implemented, ASME NQA-1, the program requirements for nuclear power plants and other nuclear facilities.[7] Compliance with NQA-1 basic requirements and supplemental requirements is one of the reasons the United States has maintained safe operations of its nuclear facilities. There have been no significant environmental impacts by U.S. nuclear power plants in part because of NQA-1 program implementation. This indicates that compliance with the E4 standard should result in similar safe operation of facilities that adopt it.

So, how does ANSI/ASQC E4 begin? Much like the others:

ANSI/ASQC E4

A primary concern of any organization or company collecting and evaluating environmental data or designing, constructing, and operating environmental technology must be the quality of the results of those programs. To be successful, an organization must have results that:

- meet a well-defined need, use, or purpose;
- satisfy customers' expectations;
- comply with applicable standards and specifications;
- comply with statutory (and other) requirements of society; and
- reflect consideration of cost and economics.

Once again, the U.S. standard contains more "legalese" than those from across the ocean. But the thrust of the four environmental management standards is quite similar, if not the same—*set up and run a management system that will maintain the existing environment as is or make it better, and accept accountability as a way of life.*

How does ISO 9000 start out? It is more like an American standard than ISO 14000 or BS 7750. Remember, ISO 9000 is on quality management. But, if we look at the opening words of its Introduction, we find amazing common ground with the environmental management standards.[8] The competitive advantage is to understand the intent of all of the standards, then tailor an organization's environmental management style and system following the guidance that best fits the situation, commitments and expectations. Never forget that sustainable development is the ultimate indicator of how you perform and how much you improve your environmental performance.

ISO 9000

Organisations—industrial, commercial, or governmental—supply products intended to satisfy customers' needs and/or requirements. Increased global competition has led to increasingly more stringent customer expectations with regard to quality. To be competitive and to maintain good economic performance, organisations/suppliers need to employ increasingly effective and efficient systems. Such systems should result in continual improvements in quality and increased satisfaction of the organization's customers and other stakeholders (employees, owners, subsuppliers, society).

So, all five standards say the same thing:

- doing business as usual is unacceptable.
- doing business without a systematic management approach is economic doom.
- providing quality is key.
- protecting and nurturing the environment is key.
- doing it right the first time is crucial.
- getting your act together is essential.
- listening to customers and other stakeholders is absolutely necessary.
- respecting society is mandatory.

And, in no uncertain terms, *pay attention* to global concerns about environmental performance and quality assurance.

The Introduction in ISO 14000 includes two sections:

- Overview
- Benefits of Having an Environmental Management System

This primer will not only reveal what's in ISO 14000, how to interpret it, and how to put it to use for a competitive advantage, but will present the fundamentals of how to design, develop, implement, and maintain an environmental management system. There will be some instructions on writing policies and procedures, and questions to answer when you start and when you evaluate the system, along with a simple flowchart of how to make the system work. This book will cover the possibilities and the problems you may encounter along the way to environmental management and sustainable development.

Major conflict is brewing between regulators and perpetrators, but between people with a common earth, a common need to sustain and thrive, and who have different motivations and priorities. Life in the coming quality environment doesn't have to be this way. If there is the cooperation expected by ISO 14000 and its companion standards, none of the combating groups need suffer. The standards are insisting that the world, through its individual human inhabitants, turn its focus toward real problems instead of petty feuds and quarrels.

Most of what we as humans do is process and consume. Most everything we get or use is from the land, usually from mining, and some things from farming or ranching. Using up resources means something needs to go back in. By committing to careful management of the quality of the environment, the world will be replenished and will continue to provide.

Without environmental responsibility, the fate of all species is questionable. A sound, well-structured, fully supported environmental management system is the key to our own survival and that of other organisms and members of the ecosystem. It'll work, if you follow the instructions of this primer. We all have everything to gain.

The protection of the environment and the health and safety of the public, and your employees, deserves as much attention to detail as protecting your personal plans for happiness and the pursuit of your ambitions. We are the family of earth. The creatures under our responsibility are both tamed and wild. Our home is global now, and becoming more and more universal as the space program expands, and we evolve.

What you do is important, perhaps greatly significant in terms of impacts on improving standards of living and preserving an environment future generations of all species will have to live in and off of. Make profits in your business or enterprise, but hold yourself accountable for wrongs against humanity and the others in our ecosystem. Unless you manage your system of environmental management, you leave all to chance and chaos. Anything can happen. Adding value is what matters. Quality is "the goal of each and every business. No company can regard quality as not a central concern and all can and must strive to attain it, maintain it, and constantly improve it" (Voehl, Jackson and Ashton 1994). This primer will guide you along the path toward quality environmental management.

0.1 Overview: Creating an Environmental Management System

The ISO 14000 document is titled *Environmental Management Systems—General Guidelines on Principles, Systems and Supporting Techniques.*[9] It outlines system elements, with advice on how to initiate, implement, improve, and sustain the system. By following its guidelines and the core elements of the environmental management system specification, ISO 14001, your organization will have a framework to balance and integrate environmental and economic interests.[10] Doing so will improve your competitive advantage.

Here's one advantage of adopting the ISO 14000 process for environmental management: The General Agreement on Tariffs and Trade (GATT) favors use of "international" standards to reduce trade barriers. International standards are said to level the playing field for more competitors. The Eco-Management and Audit Scheme regulated in the European Community should be satisfied with organizations who have an ISO 14000 environmental management system, and that can mean markets and money for a company while it fosters sustainable development. U.S. agencies favoring ISO 14000 as a voluntary standard include the U.S. Environmental Protection Agency (EPA), the U.S. Department of Energy (DOE), U.S. Department of Commerce, and U.S. Department of Justice. In fact, EPA has been a very active participant in drafting the standards.

In fact, the U.S. Office of Management and Budget also encourages the use of international standards and suggests they "should be considered in procurement and regulatory applications...." (U.S. OMB Circular A-119, Federal Participation in the Development and Use of Voluntary Standards 1993).

The ISO 14000 series consists of standards on environmental management systems. This series also covers environmental auditing, audit procedures, auditor criteria, audit management, initial environmental reviews, environmental site assessments, environmental labeling, performance evaluation, and life-cycle analysis. Other standards are in draft or yet-to-be-drafted.

Note that auditing gets a lot of attention in the series. There are environmental management system audits, compliance audits, and audits of environmental statements. In addition, there are review guidelines and guidelines on assessment, evaluation, and analysis. The gamut of environmental management is covered in fewer pages of text than one might expect. The February 1995 draft of ISO 14000 is less than 50 pages. The draft of ISO 14001 is less than half of that.

Use of ISO 14000 guidance is for organizations who want to initiate or improve their environmental management systems. Use of ISO 14001 is intended for registration bodies and for third party certification.

Third party certification involves an evaluation by the certification body, with the supplier agreeing to maintain its environmental management system for all customers unless otherwise specified in an individual contract. Third party certification can mean fewer or smaller assessments by customers. The primary disadvantage is the cost of certification and subsequent third party evaluations.

Use of both documents may be suitable for some second party recognition between contracting parties.

A customer may be interested in only certain elements of an organization's environmental management system, perhaps those that affect the ability to consistently produce product to requirements and the associated risks of production. In this case, the customer will contractually specify conformance to ISO 14000.

When second-party approval or registration is involved, an environmental management system is assessed by the customer against ISO 14001.

The choice of which standard to use depends on the organization's policy, its level of maturity (whether there is a systematic management already in place that can facilitate the introduction of environmental management), advantages and disadvantages influenced by market position or existing reputation or external relations, and the size of the organization.

Organizations covered by ISO 14000 include any "company, operation, firm, enterprise, institution or association, or part thereof, whether incorporated or not, public or private, that has its own functions and administration." So, one or more people in any line of work whose "activities, products, or services are likely to interact with the environment," (i.e., environmental aspects), should take a hard look at the benefits of implementing an ISO 14000 environmental management system.

Remember that environmental impact is "any change to the environment, whether adverse or beneficial...." If the organization has environmental aspects, it definitely has environmental impact.

A definition of the environment is in order. According to ISO 14000, the environment is the "surroundings in which an organization operates, including air, water, land, natural resources, flora, fauna, humans, and their interrelation." The environment "extends from within an organization to the global system."

ISO 14000 stipulates a set of ten management principles for organizations considering an environmental management system as follows:

1. Recognize that environmental management is one of the highest priorities of any organization.

2. Establish and maintain communications with both internal and external interested parties.

3. Determine legislative requirements and those environmental aspects associated with your activities, products, and services.

4. Develop commitment by everyone in the organization to environmental protection and clearly assign responsibilities and accountability.

5. Promote environmental planning throughout the life cycle of the product and the process.

6. Establish a management discipline for achieving targeted performance.

7. Provide the right resources and sufficient training to achieve performance targets.

8. Evaluate performance against policy, environmental objectives and targets, and make improvements wherever possible.

9. Establish a process to review, monitor, and audit the environmental management system to identify opportunities for improvement in performance.

10. Encourage vendors to also establish environmental management systems.

First, environmental management must be one of your highest priorities. Next, internal and external lines of communication must be provided. Learn what the regulatory drivers are and what possible environmental impacts your activities, products, or services may have. Obtain commitment from everyone and assign clear responsibilities, with equivalent accountability. Environmental management must occur for the life of your products.

A rigor of discipline for meeting environmental performance targets must be established, and training and resources should be readily available. Then, periodically evaluate performance by reviews, monitoring, and audits. Identify any and every opportunity for improved performance.

Encourage suppliers and subcontractors to perform in a similar fashion and to have equivalent environmental management systems.

Each step is simple and will pay for its cost in a number of ways. Knowledge is power. If these principles become part of your management style and business processes, the environment and sustainable development become an ordinary way of doing business. They become a high priority, and more so every working day. No question that regulation means liability. Do you believe anyone really cares what you do? Just ask your shareholders and those shy but powerful interested parties.

In the February 1995 draft of ISO 14000, there is an appendix of "examples" of international environmental guiding principles, formal declarations that provide a foundation for action. These declarations were the outcome of the U.N. Conference on Environment and Development held at Rio de Janeiro, Brazil, in the late spring of 1992.

This was the Earth Summit, attended by representatives from around the world. The conference was held to reaffirm the Declaration of the U.N. Conference on the Human Environment, adopted at Stockholm some 20 years prior. The concept of sustainable development is discussed in sufficient detail to provide the framework of what an environmental management system is to accomplish.

An introduction to ISO 14000 should begin with the Rio Declaration, also known as *Agenda 21*. It is the best international example of what we must do to preserve and protect our environs everywhere.

The Rio Declaration echoes Thomas Jefferson's management theories—that all men are created equal, with certain inalienable rights. Sustainable development is certainly what he had in mind, as did the other founders of the Constitution.

Sustainable development is the first principle of the Rio Declaration. The goal of the United Nations is to establish a new and equitable global partnership by creating new levels of cooperation among nations, societies, and people, with respect to the interests of all, to protect the global environment and sustainable development systems. The Declaration states that

> **Human beings are at the center of concerns for sustainable development. They are entitled to a healthy and productive life in harmony with nature.**

The United Nations believes that each nation has the sovereign right to exploit its own resources in accordance with its own environmental and developmental policies, but that each nation has the responsibility to ensure its activities cause no harm to the environments of others beyond its national jurisdiction. The Declaration further states

> **The right to [sustainable] development must be fulfilled so as to equitably meet developmental and environmental needs of present and future generations.**

Environmental protection must be an integral part of the sustainable development process. All nations are responsible for cooperating in eradicating poverty, decreasing disparities in standards of living, and meeting the needs of the majority of the world's people. A global partnership is imperative to conserve, protect, and restore the health and integrity of the earth's ecosystem.

Nations must reduce and eliminate unsustainable patterns of both production and consumption. Each nation must develop and promulgate effective environmental legislation, with environmental measures addressing transboundary and global environmental problems based on international consensus. This legislation must consider liability and compensation for victims of environmental damage expeditiously.

We all have the right to exploit our environment if we are responsible for its protection; we each have the opportunity to utilize our environment as long as we recognize our obligation to improve it for the future of civilization. Implementing an environmental management system in a manner similar to that proposed by ISO 14000 will ensure our rights, our responsibilities, our opportunities and our obligations.

Every right implies a responsibility, every opportunity an obligation.

—John D. Rockefeller Jr.

The Rio Declaration does something very unique. Principle 20 assigns a vital role to women in environmental management and sustainable development, requiring their full participation, and Principle 21 strongly recommends that the youth of the world be mobilized to forge the global partnership for sustainable development.

No longer are the roles and responsibilities given primarily to men, but to women and children. Their involvement and participation will no doubt enhance the accomplishments we can make managing, preserving, and improving the earth.

Peace is indivisible from environmental protection, says the Rio Declaration. Warfare is inherently destructive of sustainable development. At all times, the natural resources of oppressed or dominated people and of occupied nations shall be protected. The Declaration's final principle, Number 27, requires all peoples to cooperate in the global partnership to develop international law for sustainable development.

Listed below are all of the Principles of the Rio Declaration. It is an excellent source of encouragement and a constant reminder of our responsibilities and obligations as keepers of the Earth. These principles are key to your competitive advantage. Understand what the customers of the world expect from you, and you understand your markets.

The United Nations Rio Declaration

❖❖❖*Sovereign Rights to Sustainable Development*❖❖❖

Human beings are at the center of concerns for sustainable development. They are entitled to a healthy and productive life in harmony with nature. States have, in accordance with the Charter of the United Nations and the principles of international law, the sovereign right to exploit their own resources pursuant to their own environmental and developmental policies, and the responsibility to ensure that activities within their jurisdiction or control do not cause damage to the environment of other States or of areas beyond the limits of national jurisdiction.

❖❖❖*Sustainable Development for the Present and Future*❖❖❖

The right to development must be fulfilled so as to equitably meet developmental and environmental needs of present.

❖❖❖*Eradicate Poverty for Sustainable Development*❖❖❖

In order to achieve sustainable development, environmental protection shall constitute an integral part of the development process and cannot be considered in isolation from it. All States and all people shall cooperage in the essential task of eradicating poverty as an indispensable requirement for sustainable development, in order to decrease the disparities in standards of living and better meet the needs of the majority of the people of the world.

❖❖❖*Global Partnership for Sustainable Development*❖❖❖

The special situation and needs of developing countries, particularly the least developed and those most environmentally vulnerable, shall be given special priority. International actions in the field of environment and development should also address the interests and needs of all countries. States shall cooperage in a spirit of global partnership to conserve, protect and restore the health and integrity of the Earth's ecosystem. In view of the different contributions to global environmental degradation, States have common but differentiated responsibilities. The developed countries acknowledge the responsibility that they bear in the international pursuit of sustainable development in view of the pressures their societies place on the global environment and of the technologies and financial resources they command.

❖❖❖*Eliminate Unsustainable Patterns*❖❖❖

To achieve sustainable development and a higher quality of life for all people, States should reduce and eliminate unsustainable patterns of production and consumption and promote appropriate demographic policies. States should cooperage to strengthen endogenous capacity-building for sustainable development by improving scientific understanding through exchanges of scientific and technological knowledge, and by enhancing the development, adaptation, diffusion, and transfer of technologies, including new and innovative technologies.

❖❖❖*Participation of All*❖❖❖

Environmental issues are best handled with the participation of all concerned citizens, at the relevant level. At the national level, each individual shall have appropriate access to information that is held by public authorities, including information on hazardous materials and activities in their communities and the opportunity to participate in decision-making processes. States shall facilitate and encourage public awareness and participation by making information widely available.

❖❖❖*Equitable Environmental Legislation*❖❖❖

States shall enact effective environmental legislation. Environmental standards, management objectives and priorities should reflect the environmental and developmental context to which they apply. Standards applied by some countries can be inappropriate and of unwarranted economic and social cost to other countries, in particular, developing countries.

❖❖❖*International Economic Cooperation*❖❖❖

States should cooperage to promote a supportive and open international economic system that would lead to economic growth and sustainable development in all countries, to better address the problems of environmental degradation. Trade policy measures for environmental purposes should not constitute a means of arbitrary or unjustifiable discrimination or a disguised restriction on international trade. Unilateral actions to deal with environmental challenges outside the jurisdiction of the importing country should be avoided. Environmental measures addressing transboundary or global environmental problems should, as far as possible, be based on an international consensus.

❖❖❖*Liability and Compensation*❖❖❖

States shall develop national law regarding liability and compensation for the victims of pollution and other environmental damage. States shall also cooperage in an expeditious and more determined manner to develop further international law regarding liability and compensation for adverse effects of environmental damage caused by activities within their jurisdiction or control to areas beyond their jurisdiction.

❖❖❖*Cooperation for the Environment and Human Health*❖❖❖

States should effectively cooperage to discourage or prevent the relocation and transfer to other States of any activities and substances that cause severe environmental degradation or are found to be harmful to human health.

❖❖❖*Precautionary Approach*❖❖❖

In order to protect the environment, the precautionary approach shall be widely applied by States according to their capabilities. Where there are threats of serious or irreversible damage, lack of full scientific certainty shall not be used as a reason for postponing cost-effective measures to prevent environmental degradation.

❖❖❖*Internal Environmental Impact Assessment*❖❖❖

National authorities should endeavor to promote the internalization of environmental costs and the use of economic instruments, taking into account the approach that the polluter should, in principle, bear the cost of pollution, with due regard to the public interest and without distorting international trade and investment. Environmental impact assessment, as a national instrument, shall be undertaken for proposed activities that are likely to have a significant adverse impact on the environment and are subject to a decision of a competent national authority.

❖❖❖*Immediate Notification and International Response*❖❖❖

States shall immediately notify other States of any natural disasters or other emergencies that are likely to produce sudden harmful effects on the environment of those States. Every effort shall be made by the international community to help States so affected. States shall provide prior and timely notification and relevant information to potentially affected States on activities that can have a significant transboundary environmental effect and shall consult with those States at an early stage and in good faith.

❖❖❖*Full Participation by Women*❖❖❖

Women have a vital role in environmental management and development. Their full participation is therefore essential to achieve sustainable development.

❖❖❖*Global Youth Partnership*❖❖❖

The creativity, ideals, and courage of the youth of the world should be mobilized to forge a global partnership in order to achieve sustainable development and ensure a better future for all.

❖❖❖*Role of the Indigenous and Protection of the Oppressed*❖❖❖

Indigenous people and their communities, and other local communities, have a vital role in environmental management and development because of their knowledge and traditional practices. States should recognize and duly support their identity, culture and interest and enable their effective participation in the achievement of sustainable development. The environment and natural resources of people under oppression, domination and occupation shall be protected.

❖❖❖*Peace, Sustainable Development, and Environmental Protection*❖❖❖

Warfare is inherently destructive of sustainable development. States shall therefore respect international law providing protection for the environment in times of armed conflict and cooperate in its further development, as necessary. Peace, development and environmental protection are interdependent and indivisible.

❖❖❖*Peaceful Resolution, Good Faith, and Partnership*❖❖❖

States shall resolve all their environmental disputes peacefully and by appropriate means in accordance with the Charter of the United Nations. States and people shall cooperage in good faith and in a spirit of partnership in the fulfillment of the principles embodied in this Declaration and in the further development of international law in the field of sustainable development.

In the Rio Declaration international actions are encouraged to address the needs and interests of all countries. Special attention is to be given to the least developed countries and others vulnerable to environmental damage. The Rio Declaration reminds developed countries of their responsibility. They must recognize the impact of their technological and financial aspects on the world's ecosystem.

Improved scientific understanding is recommended. So is sharing new, innovative technologies for sustainable development.

All citizens with concern for the global environment are to participate in the deployment of systems for sustainable development. Citizens everywhere are to have access to pertinent information so they fully understand what hazards threaten their individual communities. Public awareness is key.

The Rio Declaration promotes an open international economic system leading to growth. It expects this international system to address environmental degradation without erecting trade barriers. Nationally legislated environmental impact assessments are to be made to provide information on potential significant adverse impacts. Otherwise, how can intelligent decisions can be made?

In the event of a natural or man-made emergency likely to harm the environment of a nation or nations, the responsible country is expected to notify them immediately, with the international consortium of countries to make every possible effort to be of aide and assistance.

The Rio Declaration, asking for the direct involvement of men, women, children, and developed and undeveloped countries, also asks for the involvement of indigenous communities to play a vital role in environmental management and development. Their culture and interests are to be respected by the nations within which they live.

The Rio Declaration is powerful statement, second only to the U.S. Declaration of Independence and U.S. Constitution. With your understanding of the Rio Declaration now ensured, we are ready to explore the one and only international standard for environmental management systems, ISO 14000.

First, let's see what guidance and practical advice ISO 14000 gives on getting management systems up and running.

A separate appendix to ISO 14000 presents the International Chamber of Commerce Business Charter for Sustainable Development. Here are eight of the key elements in their charter.

1. **Corporate Priority**. To recognize environmental management as among the highest corporate priorities and as a key determinant to sustainable development: To establish policies, programs and practices for conducting operations in an environmentally sound manner.

2. **Integrated Management**. To integrate these policies, programs and practices fully into each business as an essential element of management in all its functions.

3. **Process of Improvement**. To continue to improve policies, programs and environmental performance, taking into account technical developments, scientific understanding, consumer needs and community expectations, with legal regulations as the starting point, and to apply the same environmental criteria internationally.

4. **Employee Education.** To educate, train and motivate employees to conduct their activities in an environmentally responsible manner.

5. **Prior Assessment.** To assess environmental impacts before starting a new activity or project and before decommissioning a facility or leaving a site.

6. **Products and Services.** To develop and provide products or services that have no undue environmental impact and are safe in their intended use, that are efficient in their consumption of energy and natural resources, and that can be recycled, reused or disposed of safely.

7. **Customer Advice.** To advise, and where relevant educate, customers, distributors and the public in the safe use, transportation, storage and disposal of products provided, and to apply similar considerations to the provisions of services.

8. **Facilities and Operations.** To develop, design and operate facilities and conduct activities taking into consideration the efficient use of energy and materials, the sustainable use of renewable resources, the minimization of adverse environmental impact and waste generation, and the safe and responsible disposal of residual wastes.

These elements of the International Chamber of Commerce Business Charter provide an excellent framework for establishing an environmental management system. They ensure, with the focus on sustainable development, that you will be more able to meet customer expectations. Thus, the system should be kept alive and growing by teaching everyone to spot a problem person or situation before any injustice happens. Employees should obtain whatever useful education they can to enhance and support the system. Once the system is up and running, what are the benefits of its operational and continual improvement? Read on.

0.2 Benefits of an Environmental Management System

What does ISO 14000 say regarding the benefits to an organization by implementing an effective environmental management system? Corporations and organizations can:

- Protect human health and the environment from the potential impacts of its activities, products, and services.
- Assist in maintaining and improving the quality of the environment.
- Meet customers' environmental expectations.
- Maintain good public and community relations.
- Satisfy investor criteria and improve access to capital.
- Provide insurance at a reasonable cost.

- Gain an enhanced image and market share.
- Satisfy vendor certification criteria.
- Improve cost control.
- Limit liabilities.
- Provide resource conservation.
- Provide effective technology development and transfer.
- Provide confidence to interested parties (and shareholders) that:
  - Policies, objectives and targets are met,
  - Emphasis is on prevention first,
  - Reasonable care and regulatory compliance regularly occur, and
  - System design includes continual improvement.

Similarly, BS 7750 introduces itself first, as a specification for the various stages of developing an environmental management system, and second, as guidance on implementation and assessment.

British Standard 7750 says it is compatible with BS 5750, the mother of ISO 9000. They take parallel approaches to achieving and demonstrating compliance with specified requirements.

The contents of BS 7750 are similar to those of ISO 14000 but with some differences. The BS 7750 Table of Contents is shown below.

Specification

1 Scope

2 Informative References (versus normative references)

3 Definitions

4 Environmental management system requirements

- 4.1 Environmental management system
- 4.2 Environmental policy
- 4.3 Organization and personnel (no similar section in ISO 14000)
- 4.4 Environmental effects (same as environmental impacts)
- 4.5 Environmental objectives and targets
- 4.6 Environmental management programs (no similar section in ISO 14000)
- 4.7 Environmental management manual and documentation (no mention of the manual in ISO 14000)
- 4.8 Operational control
- 4.9 Environmental management records
- 4.10 Environmental management audits
- 4.11 Environmental management reviews

Annexes

- A (Informative) Guide to environmental management system requirements
- A.1 Environmental management system
- A.2 Environmental policy
- A.3 Organization and personnel
- A.4 Environmental effects
- A.5 Environmental objectives and targets
- A.6 Environmental management programs
- A.7 Environmental management manual and documentation
- A.8 Operational control
- A.9 Environmental management records
- A.10 Environmental management audits
- A.11 Environmental management reviews
- B (Informative) Links to BS 7750 *Quality Systems*

Table B.1 Link to BS 5750

Figure 1 Schematic diagram of the stages in the implementation of an environmental management system

List of references

ISO 9000, on the other hand, recommends embracing an environmental management system because it is:

- management motivated and
- stakeholder motivated.

Under the management-motivated approach, management initiates the effort, anticipating emerging marketplace needs or trends. Usually, a system implemented under the auspices of such motivation is generally more comprehensive and fruitful.

In the stakeholder-motivated approach, an environmental system is initially implemented in response to demands by customers or other stakeholders. Almost invariably, there will be significant improvements in internal operating results and costs.

Environmental management is very stakeholder oriented. The motivators include customers, regulators, competitors, stockholders, employees, and society. Any impact, positive or adverse, that you make on the environment needs to be by your decision.

Managers are responsible and accountable for managing corporate affairs to add value to products and operations. The ISO 9000 standard suggests four situations when quality management systems are to be implemented and two reasons why they should be implemented. The recommendations serve as guidance on how to justify an environmental management system following ISO 14000. First, the four situations:

Quality management systems should be implemented

- **as guidance for environmental management,**
- **for second-party approval,**
- **to meet a contractual agreement between first and second parties, and**
- **for third-party certification/registration.**

In a recent "Business Insight" column, it was said that the U.S. Department of Justice "'says it may choose not to prosecute a company with an [environmental management system] if it strays from compliance.' Proactive management likely reduces the number of fines for noncompliance and lowers remediation costs" (Kirschner 1995).

The article suggests that "if the history of ISO 9000 is any indication, ISO 14000 registration will become the benchmark" for environmental management systems. It quotes Ira Feldman, Vice President of Capital Environmental, a Washington, DC consulting firm, as saying, "It is global, it is voluntary, it is a consensus. It is going to become a de factor requirement of doing business voluntarily" (Ibid.).

Joe Cascio, of IBM (Somers, New York) and the Chair of the Environmental Standards writing committee, says the standards are being written with each country's current environmental progress in mind. The goal of ISO 14000 is to "bring up" companies' "environmental floors" (ASQC *Quality Progress* 1995). However, not all countries are going to have to perform at the same levels.

Cascio is also the Chair of the U.S. Technical Advisory Group to ISO/Technical Committee 207. In the same ASQC *Quality Progress* article, he describes the foreseen benefits of ISO 14000. They

> will provide a worldwide focus on environmental management; promote a voluntary consensus standards approach; harmonize national rules, labels, and methods by minimizing trade barriers and complications and promoting predictability and consistency; and demonstrate commitment to maintaining and moving beyond regulatory environmental performance compliance (Ibid.).

Cascio predicts the impact of ISO 14000 as follows: "People will be shocked at its influence, and big companies will want their suppliers to be registered to make sure they are environmentally responsible" (Ibid.). The final draft of ISO 14000 is expected to be out in mid-1996 and will be adaptable for third-party verification and self-declaration, according to the ASQC *Quality Progress* article. Certain ISO Committee Drafts are available through ASQC. They are identified in the discussion on Scope later in this primer. Casico has stressed that IBM is very much behind ISO 14000. ISO 14000 is an inevitable international standard, perhaps longer overdue than anyone realizes. Most foreign nationals have nothing like the U.S. Environmental Protection Agency and, hence, less (if any) regulatory clout on environmental insults. Information on the ISO 14000 series can be obtained from the Energy and Environmental Division, ASQC Headquarters (800) 248-1946. It is the intent of the Energy and Environmental Division to offer training and clarification support on ISO 14000. The Division last year published the ANSI/ASQC E4 standard. ASQC is a strong nonpolitical professional organization with a cadre of thousands of environmental and quality management engineers, scientists, regulators, advocates, administrators and the like.

"The ultimate aim [of developing the ISO 14000 series] is, largely, to reduce environmental waste by manufacturers and

Sidebar: Plan Ahead

The world's first uranium enrichment facility in Oak Ridge, TN, Lockheed Martin's K-25 Plant, ceased operation years ago, when the gaseous diffusion plants in Paducah, Kentucky, and Piketon, Ohio, came on line, early in the 1950s. Consider the potential environmental insults of K-25, a 1940s. vintage facility when the war took precedence over everything? It will be no easy task to return the land where K-25 is located to its former pristine state, but it will be done. Chances are, the plant will be held in a long-term standby state, while technology finds more efficient, more effective ways to decontaminate and decommission the K-25 structures, systems, and components. First, it's going to cost billions. Something that may be best passed on to future generations to figure out how to and maybe find new ways of making a profit in the effort. Something for you to consider as well. What will be the costs and benefits of decontaminating or decommissioning the facilities of your organization? Is it a better idea to sit and wait until some young blood scientist or technological entrepreneur comes along with an idea on how to profit from your demise? If you can handle the environmental impacts you have and hold them till technology catches up with you, why not? Never try to tackle a problem before you have a solution in mind, or you'll end up tackled yourself.

service providers" (Perrone and Kirkpatrick 1995). According to Perrone and Kirkpatrick,

> the market-driven forces that pushed ISO 9000 from a voluntary system to a 'de factor minimum requirement' may reappear once the ISO 14000 standards come into formal existence...This means that companies should already be planning how to address whatever environmental harm may be caused by any of their activities anywhere in the world (Ibid.).
>
> In addition to the desire to address environmental issues pro-actively through voluntary standards, and to meet consumer concerns about the environment, one of the most critical efforts is to avoid trade bottlenecks caused by differing national standards (Ibid).

Louis Jourdan, director of environmental affairs at the European Chemical Industry Council in Cefic, Brussels, believes that bankers and insurers are driving the standards (Samdani, Moore and Ondrey 1995). They need to know companies' environmental performance record to make sound lending and liability decisions. Philip Marcus, program director at Environmental Resources Management, Inc., in Exton, Pennsylvania, believes the first ISO 14000 certification could be received within a year (Ibid.).

Focus magazine warns that a growing number of companies worldwide view ISO 9000 accreditation as a prerequisite for entering new business agreements (*Focus*, Vol. 11 No. 6). The ISO 14000 standard is expected to follow suit and become the standard by which all companies in the environmental business will be measured (Ibid.).

The current International Chair of the ISO Technical Committee responsible for ISO 14000, George Connell of Canada, claims "It's not yet possible, I think, to put together a design for a general management system that will suit all corporations because they are so different...But I think any corporation can draw on ISO 9000 and ISO 14000, along with their existing knowledge of financial management and occupational safety and health management—in fact, every aspect of operations within a corporation" *(Quality Systems Update 1995).* Connell believes even the World Trade Organization will recognize the importance of environmental management systems in promoting international trade. He expects the world to turn to ISO 14000 as a solution (Ibid.).

In the same issue of the magazine appears an article on an April 1995 survey of interest (Hadlet 1995). The primary purpose of the survey was to collect market information to determine the level of awareness and plans of its ISO 9000 registered organizations with respect to ISO 14000.

Five hundred U.S. ISO 9000 registered organizations were sent questionnaires along with a newsletter on environmental management systems. The newsletter gave an overview of ISO 14000 and other environmental management standards and discussed some of the issues surrounding their development and application. The response to the survey was fairly good, with 123 completed questionnaires received (about 25%). In several instances, multiple organizations within a single company combined their response. So, actually, the response rate was closer to 40%. Only three of the organizations asked to remain anonymous. The survey found the level of awareness about ISO 14000 relatively low—64 % of respondents were either unaware or had little awareness of the development of the new standards. When questioned whether the organizations had a formal system in place to management environmental regulatory compliance, 77 % said "yes." But almost half indicated their environmental management systems were far less formal than the quality management systems.

Twenty-seven respondents indicated they had already commenced environmental management system implementation or planned to within a year. Thirty-nine percent said they planned to commence with a year to three years. And 34 % responded with a "wait-and-see" attitude. The survey found little interest in ISO 14000 from the computer software industry, based on participation by three software companies.

The survey showed most companies were driven to implementing an environmental management system by senior management motivation. Factors influencing senior management were pressure from shareholders, pressure from regulators, and pursuit of market leadership. Other factors of influence included customer pressure and realization of internal benefits.

ENDNOTES

1. The International Organisation for Standardisation was founded in 1946, according to authors James Stewart, Peter Mauch, and Frank Straka in *The 90-Day ISO Manual: the Basics*. The function of this new agency is to promote development (and now deployment) of international standards.

2. British Standard BS 7750: 1994, *Specification for Environmental Management Systems.*

3. ANSI/ASQC E4-1994, *Specifications and Guidelines for Quality Systems for Environmental Data Collection and Environmental Technology Programs.*

4. ISO/TC/207 SC1/WG2 Committee Draft of February 1995.

5. Known as DIS, Draft International Standard, out for review by countries on the Working Group responsible for it.

6. The June 1995 meeting of ISO Technical Committee 207.

7. ASME NQA-1-1989, *Quality Assurance Program Requirements for Nuclear Facilities.*

8. ANSI/ASQC Q9000-1-1994, *Quality Management and Quality Assurance Standards—Guidelines for Selection and Use.*

9. ISO 14000: 199X, ISO/TC/207 SC1/WG2 N80 rev, Committee Draft prepared by ISO/TC207/SC1/WG2; February, 1995.

10. ISO 14001:199X, *Environmental Management Systems—Specification with Guidance on Use*, Committee Draft prepared by ISO/TC207/SC1/WG1; February, 1995.

11. Pete Hunter of MAC Technical Services Company, supporting the U.S. Department of Energy in environmental management, suggests that stakeholder motivation makes the most sense, but that management motivation is usually the prime driver. "You do what your boss tells you, to motivate the stakeholders in your household. That's the bottom line, money and sustainable development. Inseparable as energy and matter."

12. Marissa Perrone is president of Eco-Trade Consulting (Philadelphia, Pennsylvania). David Kirkpatrick is in senior management for Environmental Management with Grant Thornton (Minneapolis, Minnesota). Both are members of the U.S. Technical Assistance Group to the ISO committee.

CHAPTER 1

The Scope of Environmental Management

The guidance of ISO 14000 on the development and implementation of environmental management systems and principles, including their coordination with other management systems, is applicable to any organization interested in having or improving an environmental management system. The guidelines are for voluntary use as an internal management tool. They build on the core elements of ISO 14001 and also include additional elements important to a comprehensive environmental management system. The scope of ISO 14001 is described later in this primer.

Similarly, the scope of BS 7750 says it specifies requirements for the development, implementation, and maintenance of environmental management systems but definitely refuses to "state specific environmental performance criteria," although it requires organizations to formulate policies and objectives. In the case of ISO 9000, it simply clarifies principle quality-related concepts and their distinctions and interrelationships, and provides guidance for selection and use of daughter standards.

British Standard 7750 says it is applicable to any organization that "wishes" to assure itself of compliance with environmental policy and "wishes" to demonstrate compliance to others.

The ANSI/ASQC E4 scope is described in two separate areas: one on purpose and content, and one on scope and field of application. As for purpose, E4 describes the minimum set of quality management specifications required to conduct programs of environmental data collection and evaluation or environmental technology, design, construction and operation. It is intended to provide the basis for preparation of a quality system that satisfies the unique mission of the organization.

As for scope and application, E4 provides the minimum criteria for quality systems for environmental management programs but is limited to data collection, evaluation and use, within the boundaries of environmental technology design, construction and operation.

The ANSI/ASQC E4 standard applies the minimum specifications and guidelines for environmental quality management systems and the specifications and guidelines for environmental data with regard to characterization and quantification of wastes, ambient conditions, ecological systems and processes, as well as performance assessments, treatability studies, and investigations.

A primary concern of any organization must be the quality of the results of its activities. An organization needs to be structured so that the technical, administrative, and human factors affecting the quality of its results are under control.

The "sphere" of an organization's influence needs to encompass the "sphere" of its concern for quality. The E4 standard recommends that an organization develop, document, and implement a management system designed to ensure the necessary control of quality. An organization's results must satisfy five criteria for success:

- meet a well-defined need, use, or purpose,
- satisfy customer expectations,
- comply with specified standards,

- comply with requirements of society, and
- reflect careful consideration of economics.

Both ANSI/ASQC E4 and the nuclear facility quality assurance criteria of the U.S. Department of Energy recommend a graded approach in controlling the quality of results.[1] That is, the more significant the risk of an activity, product, or service to the health and safety of the public or on the environment, the stricter the controls. Usually, a double contingency is in order, so that the failure of one component in a management system will be handled by a backup system to prevent or mitigate the consequences. Less significant risks require less stringent controls.

The scope of Quality Management and Quality Assurance Standard ISO 9000 is to clarify principal quality-related concepts and the distinctions and interrelationships between them. It also provides guidance on selecting and using the ISO 9000 family of international standards. The ANSI/ASQC Q9000-1 standard also provides guidance on American national standards on quality management and quality assurance.[2]

Scope is reach. What is the breadth and depth? Do you want to apply your environmental management to everything your organization does *or* only to specific activities, products, and services? Are there regulations that mandate the scope of your management system? Are there customers that dictate it? Think it over and think it out. Know the lengths of your arms before you try to embrace a problem or an opportunity.

ENDNOTES

1. Quality assurance criteria for the nuclear facilities in the Department of Energy complex are provided in Title 10 Code of Federal Regulations part 830, subpart 120. They are based in part on ISO 9000.

2. The ISO Technical Management Board is thinking about a standard for occupational health and safety. The United Kingdom has one, BS 8750, and other countries are developing similar standards. The ASQC Energy and Environmental Division discussed the issue in September 1995 at their conference in Charlotte, North Carolina. Occupational health and safety is closely tied to environmental management as well as quality management.

Chapter 2

Environmental Management References

Certain standards are recognized by ISO 14000 as containing provisions that are part of the standard by reference. All standards are subject to revision, of course, and readers are encouraged to check on the most recent editions before making contractual agreements. Organizations should contact members of the International European Community or ISO who maintain registers of currently valid standards.[1]

- ISO 14000 — *Environmental management systems. General guidelines on principles, systems, and supporting techniques.*
- ISO 14001 — *Environmental management systems. Specification with guidance for use.*
- ISO 14004 — *Environmental management systems. General guidelines on principles, systems, and supporting techniques.*
- ISO 14010 — *Guidelines for environmental auditing. General principles of environmental auditing.*

- ISO 14011/1 *Guidelines for environmental auditing. Audit procedures—Part 1: Auditing of environmental management systems.*
- ISO 14012 *Guidelines for environmental auditing. Qualification criteria for environmental auditors.*
- ISO 14013 *Guidelines for environmental auditing. Management of environmental management system audit programs.*
- ISO 14014 *Guidelines for initial environmental reviews.*
- ISO 14015 *Guidelines for environmental site assessments.*
- ISO 14020 *Environmental labeling. Principles of all environmental labeling.*
- ISO 14021 *Environmental labeling. Self declaration, environmental claims—terms and definitions.*
- ISO 14022 *Environmental labeling. Symbols.*
- ISO 14023 *Environmental labeling. Testing and verification methodologies.*
- ISO 14024 *Environmental labeling. Practitioner programs, guiding principles, practices and certification procedures of multiple criteria.*
- ISO 14030 *Environmental performance evaluation.*
- ISO 14031 *Evaluation of the environmental performance of the management system and its relationship to the environment.*
- ISO 14040 *Environmental management—life cycle assessment—principles and guidelines.*

- ISO 14041 *Environmental management—life cycle assessment—Goal definition/Scope and inventory analysis.*
- ISO 14042 *Environmental management—life cycle assessment. Impact assessment.*
- ISO 14043 *Environmental management—life cycle assessment. Improvement assessment (or evaluation and interpretation).*
- ISO 14050 *Terms and definitions.*
- ISO 14060 *Guide for the inclusion of environmental aspects in product standards.*

Additional information on any of the standards discussed in this book or on the status of draft standards may be obtained from the organizations listed below. It is understood that ISO 14013 through ISO 14015 are on hold until the other auditing standards are approved and issued.

ASQC[2,3]
611 East Wisconsin Avenue
Milwaukee, WI 53201
or P.O. Box 3005, 53202-3005
Attn: Customer Service
Telephone: (800) 248-1946
or (414) 272-8575
Fax: (414) 272-1734

ANSI[4]
11 West 42nd Street, 13th Floor
New York, NY 10036
Attn: Steven Cornish
Telephone: (212) 642-4969
Fax: (212) 302-1286

ASTM[5]
1916 Race Street
Philadelphia, PA 19103
Attn: Rose Tomasello
Telephone: (215) 299-5400
Fax: (215) 299-2630

Single Internal Market Information Service
Office of EC Affairs
U.S. Department of Commerce, Room 1036
14th and Constitution Avenue, N.W.
Washington, DC 20230
Telephone: (202) 377-5276

Other resources appear in the endnotes and under the Bibliography and Resource List. The normative references of ISO 14001 are provided in a later section of this primer.

Since the ISO 14000 Technical Committee and the one for ISO 9000 have agreed to share generic management elements, the ISO 9000 series references identified later in this primer are important to remember.

If and when you buy into ISO 14000, you need to remain loyal. Wavering from your commitment to environmental management using a system designed against the standard will lose you customers, favor in the eyes of your regulators, and certainly make your stockholders and employees suspicious.

ENDNOTES

1. According to the American Society for Testing and Materials, ISO 14000, 14001, and 14010 through 14012 are now Draft International Standards; ISO 14021, 14024, 14040 and 14050 are still Committee Drafts.

2. Formerly known as the American Society for Quality Control.

3. Photocopies of committee drafts of ISO 14000 are available from ASQC. They are numbered BSR14000 and BSR14001. Auditing drafts are BSR14010 through BSR14012. ASQC Customer Service has five Draft International Standards available for purchase by members and non-members.

4. The American National Standards Institute.

5. The American Society for Testing and Materials.

CHAPTER 3

Environmental Management System Definitions

A number of these definitions were presented earlier in this book, and others are referenced later on. Complete definitions found in ISO 14000 are provided here following the format of ISO 14000.

Continual improvement—process of enhancing the environmental management system, with the purpose of achieving improvements in overall environmental performance, not necessarily in all areas of activity simultaneously, resulting from continuous efforts to improve in line with an organization's environmental policy.

British Standard 7750 defines **continual improvement** as a "year-on-year enhancement" based on developments in products, services, processes, and facilities; improved product quality, operational efficiency, and resource utilization; and applications to reduce adverse environmental effects to levels equal to or less than those economically viable and of the best available technology.

Environment—surroundings in which an organization operates, including air, water, land, natural resources, flora, fauna, humans, and their interrelation. The environment in this contest extends from within an organization to the global system.

The ANSI/ASQC E4 standard defines a **customer** as any individual or organization for whom items or services are furnished or work performed in response to defined requirements and expectations. Also known as **participant** or **user**, meaning an organization, group, or individual that takes part in the planning and design process and provides special knowledge or skills to enable the planning and design process to meet its objectives, or who utilizes the results or products from environmental programs.

This standard also provides two other definitions: **demonstrated capability**—the capability to meet procurement specifications through evidence presented by the supplier to substantiate its claims and in a manner defined by the customer; **document**—any written or pictorial information describing, defining, specifying, reporting, or certifying activities, requirements, procedures, or results.

It is worthwhile to know at least two definitions found in the ISO 9000 series. They provide more depth to understanding the true scope and purpose of a management system:

contract/accepted order—agreed requirements between a supplier and customer transmitted by any means.

customer—the recipient of a product provided by the supplier.

Additional definitions of importance are provided in BS 7750:

environmental effects evaluation—a documented evaluation of the environmental significance of the effects of the organization's activities, products, and services, existing and planned.

environmental effects register—a list of significant environmental effects, known or suspected, of the activities, products, and services of the organization.

Environmental aspect—elements of an organization's activities, products and services which are likely to interact with the environment.

British Standard 7750 has these definitions of value:

environmental management manual—the documentation describing the overall system, and making reference to the procedures for implementing the organization's environmental management program.

environmental management program—a description of the means of achieving environmental objectives and targets.

environmental management review—the formal evaluation by management of the status and adequacy of the organization's environmental policy, systems and procedures in relation to environmental issues, regulations, and changing circumstances.

Environmental impact—any change to the environment, whether adverse or beneficial, wholly or partially resulting from an organization's activities products and services.[1]

Environmental management—parts of the overall management function of an organization that develop, implement, achieve, review, and maintain the environmental policy.

Environmental management system—organizational structure, responsibilities, practices, procedures, processes, and resources for implementing and maintaining environmental management.[2]

Environmental management system audit—systematic and documented verification process to objectively obtain and evaluate evidence to determine whether an organization's environmental management system conforms to the EMS audit criteria.

Environmental objectives—overall environmental goals, arising from the environmental policy and significant impacts, that an organization sets itself to achieve, and which are quantified wherever practicable.

Environmental performance—measurable outputs of the environmental management system, relating to the organization's control of the impact of its activities, products and services on the environment, based on its environmental policy, objectives and targets.

The ANSI/ASQC E4 standard gives a definition of **graded approach**—the process of basing the level of application of managerial controls applied to an item or work according to the intended use of the results and the degree of confidence needed in the quality of the results.

Environmental policy—statement by the organization of its intentions and principles in relation to its overall environmental performance which provides a framework for action and for the setting of its environmental objectives and targets.

Environmental target—detailed performance requirement, quantified wherever practicable, applicable to the organization or parts thereof, that arise from the environmental objectives and that need to be set and met in order to achieve those objectives.[3]

Interested party—individual or group concerned with or affected by the performance of an organization.

British Standard 7750 includes parties exercising statutory environmental control over the organization, local residents, the organization's workforce, investors, insurers, customers, consumers, environmental interest groups, and the general public.

management—No definition in ISO 14000. Management is defined by the manager one way and the managed, another. For example, Ron Fitzgerald, past Chair of the ASQC Energy and Environmental Division, considers it to involve vision, strategy and motivation.

It is important to define what is management and what is line function early in your development of policy and a management system. Is there a distinction between leader and manager? Manager and worker? Worker and accountability? Think about it.

Organization—company, operation, firm, enterprise, institution, or association, or part thereof, whether incorporated or not, public or private, that has its own functions and administration.

British Standard 7750 includes government departments, charities and societies as well.

Objective evidence—information which can be proved true, based on facts obtained through observation, measurement, test, or other means.

Prevention of pollution—defined in the discussion on ISO 14001 later in this primer.

Processed material—tangible product generated by transforming raw material into a desired state, typically delivered in drums, bags, tanks, cylinders, cans, pipelines, or rolls.

Product—result of activities or processes, such as service, hardware, processed materials, or software, tangible or intangible.[4]

Quality—the totality of features and characteristics of a product or service that bear on its ability to satisfy stated or implied needs.

Quality assurance—all those planned and systematic actions necessary to provide adequate confidence that a product or service will satisfy given requirements for quality.

> The definition of **quality assurance** also includes all the planned and systematic activities implemented within the quality system and demonstrated as needed, to provide adequate confidence that an entity will fulfill requirements for quality nonconformity or the nonfulfillment of intended usage techniques.

Quality policy—the overall quality intentions and direction of an organization as regards quality, as formally expressed by top management.

Requirements of society—obligations resulting from laws, regulations, rules, codes, statutes, and other considerations.

Software—an intellectual creation consisting of information expressed through supporting medium, in the form of concepts, transactions, procedures or computer programs.

Stakeholder expectations—product quality for customers; career and work satisfaction for employees; investment performance for owners; continuing business opportunity for subsuppliers; and responsible stewardship for society.

Tender—offer made by a supplier in response to an invitation to satisfy a contract award to provide product.

TQM—a management approach of an organization centered on quality, based on the participation of all its members and aiming at long-term success through customer satisfaction and benefits to the members of the organization and to society.

If any of these definitions are new to your organization or your business philosophy, it would be worth your while to create a Glossary of words, terms, definitions, acronyms, and accepted abbreviations—an internal dictionary, so to speak. Then, refer people to the glossary in your procedures. But do define unique important words or terms in procedures when their understanding is vital to accomplishing the task instructions. Be aware that a word may mean one thing to you and something totally different to somebody else. For instance, "NA" may mean "Not Applicable" or "Not Approved" or "Not Authorized." "TOC" often means Table of Contents; however, in an auditing environment, it could mean "Temporary Operating Condition."

British Standard 7750 has one more definition of merit:

verification activities—all inspection, text, and monitoring work related to environmental management.

ANSI/ASQC E4 defines **work** as—the process of performing a defined task or activity.

Other definitions from E4 include terms from the U.S. nuclear power industry:

shall—when the element is required and deviation from specification will constitute nonconformance

must—same as shall

should—when the element is recommended

may—when the element is optional/discretionary

Carelessness in the use of terms often occurs with initials. Often, the initials of an individual are sufficient evidence of accomplishing a task or review and approval. But there needs to be a document that relates the initials to the individual's name so that the authority to sign off can be demonstrated. And when initials (or names, for that matter) are documented, there should be a date, so the authority of the individual can be verified and the individual's qualifications validated.

In environmental management, as with anything else, know the meaning of the words you use and how the words are received.

ENDNOTES

1. BS 7750 calls these "environmental effects."
2. The same words are in BS 7750's definition of an environmental management system.
3. Ibid.
4. The term "product" applies to the intended product offering, not the unintended "by-products" that may affect the environment.

CHAPTER 4

Principles and Elements for Successful Environmental Management

There are five environmental management system principles within ISO 14000:[1]

❖❖❖Commitment and Policy❖❖❖

Principle 1

An organization should focus on what needs to be done—it should ensure commitment to the environmental management system and define its policy.

❖❖❖Planning❖❖❖

Principle 2

An organization should formulate a plan to fulfill its environmental policy.

❖❖❖Implementation❖❖❖

Principle 3

For effective implementation an organization should develop the capabilities and support mechanisms necessary to achieve its environmental policy, objectives, and targets.

❖❖❖Measurement and Evaluation❖❖❖

Principle 4

An organization should measure, monitor, and evaluate its environmental performance.

❖❖❖Review and Improvement❖❖❖

Principle 5

An organization should review and continually improve its environmental management system, with the objective of improving its overall environmental performance.

An environmental management system following these five principles provides order and consistency in addressing environmental concerns. Organizations *need* environmental management systems in order to anticipate and meet environmental performance expectations and ensure compliance with requirements, both nationally and internationally.

Environmental management is an essential, integral part of the overall management system. The design of the system must be an ongoing, interactive process for defining, documenting, and continually improving the required capabilities.

The same precepts appear in ISO 9000. In the words of ISO 14000,

> **An organization should implement an effective environmental management system in order to protect human health and the environment from the potential impacts of its activities, products and services; and to assist in maintaining and improving the quality of the environment.**

The Standard is especially tailored for small and medium-sized companies, although its guidance and elements of management can be used by organizations of any size. The environmental management system of any organization will provide more confidence to customers, the public, and governments that environmental objectives and targets are met. Emphasis is on prevention first, evidence of regulatory compliance and continual improvement.

The system is a framework to allow organizations to link environmental objectives with targets and with specific financial outcomes;

the tie between *eco*-protection and *eco*-nomics is indisputable, so that needed resources are provided for the best benefit to all three.

British Standard 7750 asserts that organizations must, if they're to be effective, conduct their business within a structured management system, integrated with overall management activity and addressing significant environmental "effects." Environmental management audits and reviews are inherent, says the standard, but they are separate parts of the system. By themselves, they provide no assurance that an organization's performance meets and will continue to meet legislative and policy requirements.

This British Standard has principles similar to those of ISO 14000.

- Establish and maintain an environmental management system with a defined and documented policy.
- Define and document responsibilities, authority, and interrelations of key personnel.
- Provide for verification resources.
- Have procedures to ensure effective communications to employees and interested parties.
- Provide proper training to employees and contractors.
- Identify, examine, and evaluate environmental impacts.
- Maintain a record of legislative, regulatory, and policy requirements.
- Specify environmental objectives and consequent targets.
- Have a program to achieve the environmental objectives and targets.
- Control documents and operations.

- Measure and test performance.
- Investigate and correct noncompliance with requirements.
- Maintain records.
- Perform environmental management audits and reviews.

Documented system procedures and instructions are required by BS 7750 as well as their effective implementation. Organizations must take into account any pertinent code of practices to which they subscribe.

Organizations are to define and document responsibilities, authority, and interrelations of key personnel. Key personnel are those individuals who manage, perform, and verify activities having a significant impact on the environment, whether real or potential. These personnel are to have sufficient organizational freedom to control further activities until an identified environmental deficiency or unsatisfactory condition is corrected. They are charged with the authority to initiate action to ensure compliance with the environmental policy and to identify any environmental problems. These responsible individuals are to initiate, recommend, or provide solutions to environmental management problems through designated channels and to verify implementation of the solutions.

Similar requirements to those in BS 7750 and ISO 14000 apply to the nuclear industry and, through the ANSI/ASQC E4 standard, to the chemical industry. The U.S. Department of Energy has promulgated Title 10 of the Code of Federal Regulations, part 830 (10CFR830). Subpart 120 specifies requirements for the management and control of quality in nuclear facilities (10CFR830.120) *(Quality Assurance for Nuclear Facilities).*

The 10CFR830.120 requirements came from a previous Department of Energy Order, Number 5700.6C, which was based largely on ISO 9000 and the "graded approach" to quality (Ibid.). The U.S. Nuclear Regulatory Commission has almost identical requirements in 10CFR50 Appendix B, showing 18 criteria for effective management and control of quality. The U.S. Department of Energy squeezed the 18 criteria into 10 to more closely align them with the respective responsible parties.

Ten management elements are identified by the ANSI/ASQC E4 standard:

1. organization
2. management system description
3. qualifications and training
4. procurement
5. documents and records
6. hardware and software
7. planning
8. work processes
9. assessments and response
10. continuous improvement

According to ANSI/ASQC E4, it's up to management to define the organization. Management must identify functions and responsibilities, levels of authority and accountability, lines of communication, and interfaces. Management identifies the needs and expectations of customers, then defines objectives to satisfy customers.

It is up to management to provide the necessary resources to the line to accomplish objectives. Management has to create a work setting that is conducive to personnel collaboration on producing the type and quality of results intended. Managers are ultimately and directly responsible for the success of the system they design.

The system must be planned, established, documented, and communicated to everyone with responsibilities. Logistics, such as when and how controls are to be applied, need to be identified. These controls are to be applied commensurate with their importance to the environment, human health and safety, and the success of meeting objectives. System implementation must continually be assessed for compatibility and consistency with activities, products, and services.

People make the system work. They need to be properly qualified, competent and further empowered with sufficient authority. This involves adequate training on requirements, techniques, the use of tools, management expectations, resource availability, and the importance of the controls. Performance and proficiency need to be evaluated periodically.

Your environmental management system must be a descendant and proponent of your management and business philosophy. And philosophy always starts with principles and ends with expected practices.

4.1 Environmental Management Commitment and Policy

Remember Principle 1? It is restated at the beginning of Section 4.1:

An organization should focus on what needs to be done—it should ensure commitment to the Environmental Management System and define its policy.

This section of ISO 14000 covers the following elements:

- Top Management Commitment and Leadership
- Initial Environmental Review
- Environmental Policy

Begin where there is recognizable benefit, says ISO 14000. Focus on:

regulatory compliance
or
limiting sources of liability
or
making more efficient use of materials

British Standard 7750 suggests an environmental management system be designed to enable the organization to maximize its beneficial effects and minimize its adverse effects, with emphasis on prevention instead of detection and amelioration after damage has occurred. It suggests that organizations identify and acquire or develop needed skills, equipment, controls, processes, monitoring systems or whatever tools are needed in order to achieve the required environmental performance.

Both BS 7750 and ISO 14000 allow for sharing common elements of the existing overall management system regarding operations, occupational health and safety, and environmental protection so that shared documentation and records will reduce duplication. In these cases, the interrelationships are to be identified and cross-referenced. Effective integration and coordination is essential to ensure consistent decision-making.

The British Standard includes a discussion of the environmental management program. Following the program is key to compliance with the environmental policy. Implementing the program means a clear and unequivocal commitment by everyone involved, especially the most senior levels of management. The most senior officer must ensure that suitable organizational systems are in place.

The program should deal with any environmental consequences from past activities of the organization and with development of new products or services throughout their life cycle—from feasibility studies through planning and design to construction, installation, operation, and eventual decommissioning.

Key objectives and responsibilities of an organization, according to ISO 9000, include

1. achieving quality in activities, products, and services,

2. maintaining quality in activities, products, and services, and

3. seeking to continuously improve quality to meet the stated and implied needs of stakeholders.

Organizations should demonstrate quality improvement and provide confidence that requirements are being fulfilled. Management is responsible for the policy, the organization, and management reviews, according to the ISO 9000 series, including objectives and commitments. Adequate resources must be provided by management to achieve the objectives and meet the commitments. Management reviews of the system must occur at defined intervals on a frequency that ensures continuing suitability and effectiveness.

Environmental management is the safest, least expensive way to make your organization survive. Don't kill the egg until the chicken has laid it. Count the eggs in the basket but only <u>after</u> you've counted the chickens in the coop. The chicken precedes the egg, to end a lifetime dilemma. It may have come from a pre-chicken source, but without the chicken....

This chicken/egg analogy is about environmental understanding. In order to ensure golden eggs for your organization and for sustainable development, you need to know the source of the egg. Are your suppliers environmentally conscious? Do your suppliers give you the quality of product or service you need to ensure the quality of your own product or services? Are you masking inferior supplied items in your final product?

By asking a few questions, you can better understand what value you are adding to your product quality, environmentally. Perhaps you are adding value that should already be there, provided by your suppliers. Although ISO 14000 has no requirement that you mandate sound environmental management by your suppliers, it strongly recommends that you encourage them to participate.

Without your personal commitment and a solid commitment from your people, the policy is only a dusty document on the shelf. It has no clout, no purpose, no one pursuing it. Policy comes from the heart of an organization, from the people who are the organization. To make yours work, you must commit to making it work.

4.1.1 Leadership in an Environmental Management System

The first step in environmental management is top management commitment to review and improve environmental performance. Step two is to provide the necessary leadership. Finally, a full, honest review of your environmental performance and performance capabilities is required.

Leadership entails taking the initiative to implement new processes. BS 7750 recommends preparing an environmental management manual to describe the system. The manual will serve as a permanent reference to the implementation and maintenance of the system. This manual covers the whole organization. It may involve subcomponents such as division-level manuals. It may require more specialized manuals for individual functions, like design, marketing, finance, and investment, or individual process lines. All the manuals should be consistent in approach and content. Every manual should be subject to similar rules for control, review and revision.

It may be wise for the site emergency plan and the occupational health and safety manuals or documents to incorporate relevant environmental information and instructions. Another key function of the environmental management manual would be for audits to verify if it fits its purpose. Procedures in the manual need to be simple, unambiguous, and understandable in the methods and criteria they prescribe.

Other important attributes of the environmental management manual and documentation include dealing with normal as well as abnormal operating conditions, incidents, accidents, and potential emergency situations. Methods to deal with emergencies need to be periodically tested for effectiveness and suitability.

Peter Block suggests that leadership means to create order; that is, consistency, control, and predictability (Block 1993). He says that control means there is a clear line of authority.

Leaders are successful when they distribute ownership and responsibility, when they balance the power and promote a partnership to end secrecy, demand a promise, and redistribute wealth.

Stephen Covey, the management "guru" for the U.S. Department of Energy, suggests similar traits and talents. He describes the habits that apparently all effective managers share (Covey 1989).

HIGHLY EFFECTIVE HABITS

1. **Be proactive**
2. **Begin with the end in mind**
3. **Put first things first**
4. **Think win/win**
5. **Seek first to understand, then to be understood**
6. **Synergize**
7. **Sharpen the Saw; principles of balanced self-renewal**

Proactivity means being responsible, with the knowledge that behavior is a function of our decisions, not our conditions. Responsibility is the ability to choose a response. It's not what happens to us but our response to what happens to us that hurts or helps us. Consider the alternatives. Choose a different approach. Control your own feelings. Create an effective presentation. Choose an appropriate response. Choose, prefer, will. Your "circle of influence" is smaller than your "circle of concern." A proactive focus and positive energy enlarge your circle of influence until it encapsulates the concern.

The power to make and keep commitments is the essence of developing the basic habits of effectiveness. Knowledge, skill, and desire are all within our control. We can work on any one to improve the balance of the three. As the area of intersection becomes larger, we more deeply internalize the principles upon which the habits are based and create the strength of character to move us in a balanced way toward increasing effectiveness in our lives. As Covey says, "Proactive people make love a verb." He also quotes Thomas Watson, the founder of IBM: "Success is on the far side of failure."

Where you're heading depends on where you're coming from. Start with a clear understanding of your destination. Know where you're going so you better understand where you are now, and the steps you take will always be in the right direction. All things are created twice: first, a mental creation; then, a physical or social creation, a blueprint, a script (my interpretation of Covey's principles). Covey believes, "Leadership is not management. Management is the second creation." Leaders ask "What do I want to accomplish?" Management asks "How should it be accomplished?"

"Management is doing things right; leadership is doing the right things," according to Peter Drucker and Warren Bennis (Ibid.). Management is efficiency in climbing the ladder of success, but leadership determines whether the ladder is leaning against the right wall.

Highly successful habits, according to Covey, require imagination and conscience.

> An organizational mission statement—one that truly reflects the deep shared vision and values of everyone within that organization—creates a great unity and tremendous

> commitment. It creates in people's hearts and minds a frame of reference, a set of criteria or guidelines, by which they will govern themselves. They don't need someone else directing, controlling, criticizing, or taking cheap shots. They have bought in to the changeless core of what the organization is about (Ibid.).

Effective management is putting first things first. Leadership decides what those first things are, and management is the discipline of carrying them out. Time management is organizing and executing around priorities. "If we delegate to time, we think efficiency; if we delegate to other people, we think effectiveness" (Ibid). Stewardship delegation involves clear, up-front mutual understanding and commitment regarding expectations in five areas: desired results, guidelines, resources, accountability, and consequences. "The key is not to prioritize what's on your schedule but to schedule your priorities" (Ibid.).

A leader creates a clear, mutual understanding of what needs to be accomplished, focusing on results, not methods. An effective leader will identify the parameters within which the organization should operate, including any formidable restrictions and failure paths. Leaders identify resources required to accomplish objectives (human, financial, technical, organizational). They set standards of performance to use in evaluating results and specify what will happen, both good and bad, as a result of the evaluation.

There are six paradigms of human interaction: win/win; lose/lose; win/lose; win; lose/win; or no deal. Character is the foundation of win/win and involves integrity, maturity, and what Covey calls the "abundance mentality." There are five elements of the win/win paradigm:[2] desired results, guidelines, resources, accountability, and consequences. First, see the problem from the other point of view. Second, identify the key issues and concerns involved. Third, determine what results would constitute a fully acceptable solution. And fourth, identify possible new options to achieve those results.

Empathetic listening is the key, says Covey. Diagnose before you prescribe. Understand and perceive. Commit, then learn, then act.

One of Covey's best known concepts is that of your circle of concern and your circle of influence. You always want your circle of influence to encompass your circle of concern, with more influence than concern.[3]

Good leaders think in every dimension they can lay their hands on. The Boy Scouts of America teach leadership skills, and the first and foremost is "Be Prepared."

The best management practices are structured around customers (Block 1993). Manage the customer relationship first and design the work flow to fit the relationship. Purchasing and supplier relationships are almost equally important. Discipline must be a standard practice. "[There are] three principles in man's being and life, the principle of thought, the principle of speech, and the principle of action. The origin of all conflict between me and my fellow men is that I do not say what I mean, and that I do not do what I say."[4]

Astute managers know customer loyalty is an absolute necessity for profitable businesses from now on (Bell 1994). Bell recommends that managers adopt the attitude of customer partnership, an orientation that starts with a deep, assertively demonstrated respect for the customer. You need to maintain a spirit of contribution and "the joy of knowing the best possible has been done to meet or exceed a need" (Ibid.). The current market demands, since the end of the 1950s, have been customer-driven. Most companies in the United States at least, are obligated to tell you where to find what you're really looking for, rather than selling you a product or service you don't need. Customers are both external and internal. Anyone you give something to is a customer. Anyone you take something from is a supplier. The effective environmental management system will keep these two concepts in mind at all times. The organizations that succeed will be "those that invite the customer to access them any time, in many ways, and with ease" (Ibid.). Here are the elements Bell identifies in his anatomy of partnership:

	Abundance
+	**Trust**
+	**Dreams**
+	**Truth**
+	**Balance**
±	**Grace**
=	**Customer partnership**

An effective leader of an environmental management system will think in terms of two economic equations (Silverstein 1993).

What's good for the environment = What's good for the economy
What's good for the economy = What's good for the environment

Silverstein writes, "The health of the world's ecosystems and the wealth of the world's economics now ebb and flow in tandem."

There are six functions in the work of a manager (Allen 1994):

1. Establishing objectives
2. Organizing
3. Motivating
4. Developing people
5. Communicating
6. Measurement and analysis

> Considering the entire firm as 'the team' and expecting synergies and interdepartmental solutions to naturally evolve is generally far too optimistic except for the most simple issues.... For whatever reason, there is no synergy. In fact the opposite results in a very expensive fashion: The whole is less than the sum of its parts. True systems improvements often result from actions in the 'white spaces' on the organizational chart." Using interdepartmental teams "to identify and eliminate chronic problems

> are about the only proven technique that can systematically solve system issues, and at the same time improve the culture to remove some departmental boundaries (Fellers 1992).

Dr. Fellers recommends that interdepartmental teams have at their use at least nine tools (Ibid.):

- Brainstorming
- Cause and effect diagrams (fishbone charts)
- Pareto charts
- Flowcharts
- Trend charts
- Control charts
- Scatter diagrams
- Other statistical tools
- Common sense

He also notes five deadly diseases for any organization that are worth repeating here:[5]

1. Management by-the-numbers
2. Performance appraisal by-the-numbers
3. Lack of constancy of purpose
4. Mobility of management
5. Short-term orientation

A simple solution is to recognize the following traits as intangible and "world class" (Ibid.):

- ✓ Creativity
- ✓ Energy
- ✓ Inquisitiveness
- ✓ Desire to work for the good of the team
- ✓ Eagerness to compensate for co-workers' shortcomings
- ✓ Willingness to achieve results that are necessary, but unnoticeable to the boss

Leadership is key to making the management system work. Leaders learn from history and continually evaluate their own unique capabilities to add to those from previous leaders. They try to avoid "reinventing the wheel" and use what is already there. But when the tried and proven techniques no longer fit, they try a new approach. If the stride of the leader before you seems wrong, you must make your own new footsteps.

4.1.2 The First Environmental Review

The initial environmental performance review should consider the full range of operating conditions, opportunities for significant environmental impacts or damages from emergencies. It is important to identify and document the areas to be reviewed, be they activities, specific operations, or specific sites. The review may be conducted by questionnaires or interviews, checklists, direct inspections and measurements, records review, and is always benchmarked by looking inside and outside the organization.

The aim of this preparatory environmental review, according to BS 7750, is to consider all aspects of the organization to identify strengths, weaknesses, risks and opportunities in four areas:

? legislative and regulatory requirements

? evaluation and documentation of significant impacts

? examination of existing practices and procedures

? assessment of feedback from incident investigations and noncompliance

Throughout ISO 14000 there are boxes inserted to offer practical help, which include information like that in the previous paragraph. The document flags a few bulleted items to consider, like the use of questionnaires and checking with enforcement agencies on laws and permits. The practical help guides of ISO 14000 are discussed where applicable throughout this book.

Remember, the goal of an environmental management system is improved performance through a structured process in light of economic and other circumstances of the organization. The system is simply a tool to achieve and systematically control environmental performance levels. The level of complexity of the system is dependent on the size of the organization and the nature of its activities, products and services. What is the minimum the system should accomplish?

- Establish appropriate environmental policy.
- Identify environmental aspects and significant environmental impacts.
- Identify relevant requirements, legislative and regulatory.
- Identify priorities, appropriate environmental objectives and targets.[6]
- Establish a structured process to implement policy, achieve objectives, and meet targets.

- Plan, control, monitor and review policy implementation for continuous improvement.
- Adapt to changing circumstances.

The initial performance review needs to keep the following goals in mind.

- ➢ What environmental expectations do you need to meet? Want to meet?
- ➢ How are your public relations? Community support and involvement?
- ➢ How can you satisfy your investors? How's your access to capital?
- ➢ Do you have enough insurance? At what cost?
- ➢ Do you need to or want to improve your image? Your market share?
- ➢ What vendor certification criteria do you need to meet?
- ➢ How's your cost control?
- ➢ Do you have adequate limitations on your liabilities?
- ➢ How do you demonstrate responsible care of the environment?
- ➢ Are you able to conserve energy? Resources?
- ➢ How are you on obtaining permits and authorizations?
- ➢ What processes are in place for technology development and transfer?
- ➢ How are your relations with your industry? With government?

Those goals are also the benefits you can expect from your review and the environmental management system you develop, document, and decide to go with. Areas recommended by ISO 14000 for consideration are:

- Identify legislative and regulatory requirements.
- Identify environmental aspects, liabilities, and significant environmental impacts.
- Evaluate and document significant environmental issues.
- Evaluate performance against relevant internal criteria, external standards, regulations, codes of practice, sets of principles and guidelines.
- Consider existing environmental management practices and procedures.
- Identify policies and procedures for procurement and contracting.
- Obtain feedback from previous incident and noncompliance investigations to identify opportunities for competitive advantage.
- Assimilate the views of all interested parties.
- Identify and evaluate other functions and activities of the organization that could enable or could impede environmental performance.

The standard suggests looking to a few outside sources during the initial review—enforcement agencies for laws and permits, libraries and databases, other organizations interested in exchanging information (also good for benchmarking), industry associations and large consumer organizations, the manufacturers of your equipment and machinery, companies who transport and dispose of your wastes and, of course, professionals, like consultants. Never forget your local library, usually an excellent source of information. Most have access to good on-line services.

Keep in mind, the first review of your activities, products, and services should be thorough. You need to know your history, your present environmental aspects and impacts, and those in the future. If information already exists, use it.

BS 7750 Tips on Areas to Review

- places or functions where environmental performance could be improved
- views of relevant interested parties
- environmental objectives and targets beyond regulatory requirements
- expected changes in regulations and legislation
- adequacy of resources and environmental information
- environmental records
- cost-benefit analyses and accounting methods
- internal and external communications
- environmental aspects
- resource consumption
- waste minimization and recycling initiatives
- use of hazardous processes
- use and disposal of hazardous materials and products
- transportation policy

- ➢ nature conservation
- ➢ complaints and follow up
- ➢ visual impacts, noises and odors
- ➢ environmental priority of suppliers
- ➢ hazards, risk assessments and potential emergency situations
- ➢ emergency planning
- ➢ investment policies

The initial environmental performance review should consider dependability characteristics of products. Dependability, according to ISO 9000, involves reliability, maintainability and availability. There are four facets of product quality to keep in mind:

- Quality due to defining and updating the product.
- Quality due to designing into the product the characteristics that enable it to meet requirements or to provide value to customers and other stakeholders.[7]
- Quality due to maintaining consistency in conforming to product design, providing the designed characteristics, and providing the designed values for customers and other stakeholders.
- Quality due to furnishing support throughout the product life cycle to provide the designed characteristics and values

All four facets contribute to product quality while product value involves both quality and price.

The Department of Energy and the Environmental Protection Agency encourage a process for collecting environmental data. It is based on having Data Quality Objectives (known as DQOs). A similar process may be of value in conducting your first environmental review.

- Understand the context of the problem.
- Identify major questions and decisions.
- Determine associated measures and data requirements.
- Identify acceptable sampling risks and error tolerances.
- Define logic for making decisions with the measures.
- Determine optimal methods for data collection, sampling and analysis.

First, state the problem, the context of the review. Make sure the context is understood. Identify key personnel, their roles and responsibilities. Identify what data will be collected and why. This works best if you identify the questions to be asked and what objective evidence will be needed for the answers (know the answers you expect or don't ask the question).

Identify the decision logic and what decisions need to be made about the data and information or resources to be input to the decisions. The decision logic should be diagrammed with an **IF...THEN** set of rules and action steps (how will data trigger a decision?). List the variables and characteristics to be measured and what information might be necessary to resolve decisions. Include the rationale for the chosen variables.

Define the boundaries of the review, both spatial and temporal, and any physical constraints. Specify acceptable limits on decision errors (tolerances). Develop sampling and analysis alternatives. Review historical data. Define, specify, develop, and review—the complete cycle.

After the initial environmental performance review is complete (or before), develop and document your vision and your mission statements. If they have already been developed, include them in the review. Your review will help you to accomplish the key steps in coming up with a workable environmental management system. Tools for the initial and other environmental reviews may include:

- ✔ questionnaires
- ✔ interviews
- ✔ checklists
- ✔ inspection and measurement
- ✔ records review
- ✔ benchmarking

Your vision must be based on your purpose and capabilities. Your mission, on your customers and their expectations. You need to know where you're coming from if you're to know where you're going, as we said earlier. Let's say you want to be the best in your industry. You have to know where you stand now and why, and about your history of operation and aggression. You also have to have a plan in mind, and that means understanding your purpose for wanting to go forward.

Charisma is "A special quality of leadership that captures the popular imagination and inspires unswerving allegiance and devotion.... A person who has some divinely inspired gift, grace, or talent" or "magnetic charm or appeal" is a charismatic leader (Richardson and Thayer 1993). Most charismatic characteristics can be learned.

Your initial environmental review should be led with charisma and hutzpah. In the beginning, charisma was used in the religious sense and meant "a speaking through one's soul, being set afire by the Holy Spirit (Ibid.). Max Weber, an early management theorist and founder of modern sociology, borrowed the word, and overnight it became a way of describing someone with capabilities less definable with other words. Richardson and Thayer suggest that one thing was obvious: "People with charisma created a strong bond with those they lead." To them, charisma means you can "move others into action." Hutzpah implies tenacity.

Management, motivation, leadership, and, yes, sustainable development, all must come from the heart. Full commitment is mandatory.

4.1.3 Writing Policy for Environmental Management

An environmental policy is a statement by an organization of its intentions and principles for environmental performance. It is the framework for action and sets environmental objectives and targets. Policy establishes a sense of direction within set parameters and aims for the "overarching" goal of environmental performance. All subsequent actions by the organization will be judged against the goal of the policy statement.

The Rio Declaration is an excellent resource for tailoring your policy to include comprehensive and organization-specific principles, objectives, and targets.

Here's what ISO 14000 says about what should be included in policy considerations:

- Your organization's vision, core values, beliefs, and mission,
- Requirements of interested parties,
- Communications with interested parties,
- Continual improvement opportunities,
- Interactive alignment with other organizational policies and elements, and

- Local and regional conditions.

The members of an organization's management with the highest proprietary interest must craft and disseminate the policy. They must recognize that any of the organization's activities, products or services may impact the environment, both negatively and positively, and *that* recognition needs to be captured in the spirit of the policy. Other commitments to be dealt with in the policy include:

✔ Minimizing environmental impacts of new developments by integrating management planning and procedures.

✔ Developing environmental performance evaluation indicators and procedures.

✔ Embodying life-cycle thinking and designing product and processes to prevent pollution, reduce waste, and consume fewer resources.[8]

✔ Ensuring adequate education and training of responsible individuals and customers.

✔ Providing technology and information transfer.

✔ Involving and communicating with interested parties inside and outside the organization.[9]

✔ Working towards sustainable development.

✔ Encouraging suppliers and contractors to have environmental management systems as well.

The environmental policy is the "driver," according to ISO 14000, "for implementing and improving the organization's environmental management system so that it can maintain and potentially improve its environmental performance." The policy reflects top management's commitment to compliance with applicable laws and regulations and to continual improvement.

Continual improvement is the process of enhancing the environmental management system. The aim is to achieve improvements in overall environmental performance from continuous efforts to improve activities simultaneously or in line with the organization's scheduled priorities.

It is the policy that is the basis for environmental objectives and targets. It should be sufficiently clear so that internal and external interested parties understand it. Policy is a living document that requires periodic review, constant nurturing, and revision when needed to reflect changes in conditions or information.

The environmental management system specification, ISO 14001, requires that the policy accomplish certain minimum actions:

- Be appropriate to the nature, scale, and environmental impacts of the organization's activities, products, and services.

- Include a commitment to continual improvement and prevention of pollution.

- Include a commitment to comply with relevant environmental legislation and regulations as well as other requirements to which the organization subscribes.

- Provide the framework for setting and reviewing environmental objectives and targets.

- Be documented, implemented, maintained and communicated to all employees.

- Be available to the public.

The British Standard 7750 stresses that the environmental policy must be in a "readily understood format."

Throughout my years of reviewing policies, it has been my observation that most people write for their boss or their boss's boss or bureaucrats who read to please their boss. Remember who you're writing for, who you're communicating with, your audience, and write in their language. Keep it short and simple.

A policy statement should resemble a poem rather than a thick, boring book. If you know of organizations that have good written policies, adopt their style or format. If you know someone in your organization who can draft environmental policy better than you can, use their talents.

If you hire an outside consultant to write policy, make sure the result is in the language of your audience. No policy has to be flat; in fact, policy should be colorful, vibrant, future-focused, motivating, and moving. It needs to reflect the way your organization does business and wants to do business.

Your organization should always be in line with your policy. When you fail to follow policy, you have a serious problem deserving immediate attention. So, make your policy preventative in nature, a guide that keeps you traveling down the right road to success in environmental protection and improvement.

The British Standard advocates that the environmental policy

- demonstrate relevance to the organization's activities, products, services, and their environmental impacts;
- be communicated, implemented, and maintained at all levels of the organization;
- be made publicly available;
- show commitment to continual improvement in environmental performance;
- provide for setting and publishing environmental objectives;
- identify what activities are covered by the environmental management system; and

- indicate how environmental objectives will be available to the public.

Although ANSI/ASQC E4 says little about policy since it has a limited scope of application, it does provide useful guidance in its discussions on specifications. According to ANSI/ASQC E4, the quality policy of an organization should ensure that environmental programs produce the required type and quality of results, both needed and expected.

Management is to identify the internal and external customers toward which the policy is directed. Policy needs to consider procurement of items and services as well, to satisfy the acceptance criteria of the customer. Management oversight and inspection of work processes need to be commensurate with the importance of the work and the intended use of the results. A sound environmental management policy will contain management and technical assessments.

Continual improvement needs to be a significant part of the policy. There should be a "no fault" attitude to encourage identification of problems. Management should encourage everyone in the organization to exceed the expectations of their customers whenever possible.

> Successful leaders get all types of people with differing backgrounds, beliefs, and values to focus on a project in total synchrony. Your policy statement should do the same. Otherwise, you will have a commitment document without committed resources. Infuse your vision into those around you" and make your vision the lightning rod for the limitless energy of human promise (Richardson and Thayer 1993).

Your policy is like a compass. "It is a prerequisite for any leader to first define a direction and make sure his or her organization has the resources to accomplish it." Human beings "are emotionally driven creatures," and people must be naturally stimulated emotionally if they are to be led into action. Ignite people's deepest capabilities and enable them to perform with excellence and dedication. If you can aim the thoughts of others, you can direct their emotional state. Guide their emotions and you guide their actions (Ibid.).

Your policy is to motivate your organization into positive actions. Consider the keys to motivation offered by Richardson and Thayer:

- Phrase things in terms of what can be done and what will be done.
- Define problems in finite terms.
- Provide a clear-cut course of action.
- Inspire with passion.
- Be sincere.
- Be precise.
- Be optimistic.
- Involve a sense of timing.
- Establish commonality.
- Know what outcome you want.
- Know what you want people to do.
- Turn ideas into actions.
- Intrigue, enthuse, and motivate.

Your policy will determine the quality of your entire environmental management system.

4.2 Environmental Management Planning

Principle 2 is restated at the beginning of Section 4.2 of ISO 14000:

> **An organization should formulate a plan to fulfill its environmental policy.**

This section provides guidance on five elements for developing an environmental management system:

- Identification of Environmental Aspects and Evaluation of Associated Environmental Impacts
- Legal Requirements
- Internal Performance Criteria
- Environmental Objectives and Targets
- Environmental Management Plans and Program

In the ANSI/ASQC E4 standard are requirements for a quality management plan document using a tiered approach. A tiered approach establishes common "umbrella" functions to serve all types of environmental programs while retaining enough flexibility to meet specific technical needs. The entire organization-wide management system should be described in a plan or manual and should be supported by applicable plans or similar documents for work activities.

To manage effectively and control your influence on the environment, work processes need to be performed according to approved, pre-planned procedures. When work is supervised directly or is within the skill of the craft and of no imminent safety concern, step-by-step procedures may be unnecessary. But such activities should be identified in the plan of action.

In planning, ANSI/ASQC E4 requires coordination among all participants. Define the scope, objectives and desired results. Identify the participants. Identify the documents or data required to achieve the results. Identify pertinent requirements and controls. Address any special skills needed. Be systematic in planning and allow for deviations to be reported to management.

Activities involving design, construction or operation all require planning. Key users and customers must be involved as well as others responsible for input or action. Planning means knowing your destination and how to determine when the acceptance criteria is satisfied. It means identifying necessary resources and having a means to allocate them effectively.

> Successful players in the global marketplace invest in the countries in which they sell because it brings them closer to customers.... Through their governments and their spending habits, consumers everywhere are forcing companies to shoulder the environmental consequences of their activities (McInerney and White 1993).

Your environmental management planning should consider this insight. Consumers "also expect the companies they buy from to meet their environmental obligations in ways that go beyond public relations" (Ibid.). After studying businesses in Japan, McInerney and White have concluded, "Japan will develop its environmental opportunities the way it attacked information technology—mobilizing government and industry from top to bottom toward a single, well-defined objective." However, their opinion of U.S. companies is far different. "In American industry, environmental responsibilities are primarily viewed as a burden, a not-so-hidden tax imposed by a vocal, granola-eating minority" (Ibid.).

Planning isn't new, nor is environmental management. McInerney and White put it well: "Concern about the environment—and wasteful consumption, for that matter—is not new. But real pressure to use our natural resources more carefully only began twenty years ago." They point to the first OPEC oil embargo of 1972. "Waiting in gas lines, everyone could consider the fragile link between industrial growth and the environment" (Ibid.). The impact of the oil embargo was immediate and severe.

McInerney and White offer other suggestions for planning an approach to environmental management. First,

> Employees don't like being kept on a short leash, and neither do customers." With regard to the other end of the production line, they suggest that an important step to better quality is to "shift the responsibility to your suppliers. But you must do more than subcontract, you must trust your suppliers (Ibid.).

Plan for failure as well as success. Plan contingency routes, alternatives, and those actions that will mitigate any failure. Document the plan and flowchart it to check for illogical steps, steps out of rhythm or out of harmony. In any plan there are at least two schedules, the critical path and the schedule for activities parallel to the critical path. Critical path is the shortest schedule from start to finish, showing those activities that must precede and follow each other. Most activities can be accomplished in parallel to the critical path.

Take building a house. First comes the excavation, then the foundation, then the structure. While you're digging, you can also place an order for future materials like paint, fixtures, windows, and so on (parallel activities). But the excavation must be finished before the foundation is placed (critical path). While laying the foundation, a driveway could be installed (parallel activity), but the foundation must be complete to erect the walls (critical path). See the difference between these paths and how to interweave both schedules in your plan?

Plan your failures as well as you do your successes. Expect calamities and calculate alternatives, contingencies. Plan mitigations to lessen the impact of failures. And learn from every failure, every near-miss, every mishap, mistake, and stumble.

4.2.1 Identifying Environmental Aspects and Evaluate Impacts

Before an organization can establish its policies and prioritize its environmental objectives and targets, according to ISO 14000, it needs to identify the environmental aspects and significant environmental impacts associated with its activities, products, and services. Ensure that the significant impacts are taken into account in setting environmental objectives and associated targets.

Identifying environmental aspects is an ongoing process. It determines the past, present, and potential positive and negative impacts on the environment. It also includes identifying potential impacts to regulatory, legal, and business exposure and may include identifying impacts on the health and safety of people or the environment.

The procedure recommended by ISO 14000 for identifying environmental aspects and evaluating potential environmental impacts is a three step process:

Step 1 Select an activity or process large enough for meaningful examination but small enough to be clearly understood; for example, product design.

Step 2 Identify as many environmental aspects as possible associated with the selected activity or process; for example, product design involves packaging, raw materials, processing energy, etc.

Step 3 Evaluate the significance of impacts. Consider both the environmental and business concerns.

Environmental management planning is one area where ISO 14000 goes beyond the guidance of BS 7750. Planning is only implied in BS 7750, with no direct instruction on what to accomplish. Organizations must plan their success to be prepared for opportunities to fail. Planning means looking ahead as far as possible, then mapping out a path to get where you want to go. You also need to know where you're coming from to know where you're headed—both in direction and your intentions.

You probably have a wealth of information available for identifying the environmental aspects of your activities, products, and services. Research the information whether the outcome is negative or positive. What is the composition of the supplies you use? What are the shelf lives? How are materials disposed of? Are there more environmentally benign materials that could be procured? What are your competitors' environmental aspects?

Environmental Concerns

? the scale of the impact

? the severity of the impact

? probability of occurrence

? permanence of impact

Business Concerns

? potential regulatory/legal exposure

? difficulty of changing the impact

? cost of changing the impact

? effect of change on other activities/processes

? concerns of interested parties

? effect on public image

According to ISO 9000, all work is accomplished by a process, with inputs and outputs. Outputs are the results of the process, the products. The process is a "transformation that adds value." Every process involves people and resources. The quality of the process depends on management of the structure and operation of the process and the quality of the product or information within that structure.

Organizations exist for one reason—to add value. Work is accomplished through a network of processes for the many functions to be performed. Organizations need to identify, organize, and manage the network of processes, interfaces, and interrelationships. It is through the management of the network of processes that an organization creates, improves, and provides consistent quality.

- Are the processes defined? Procedures documented?
- Are the processes fully deployed? Implemented as documented?
- Are the processes effective? Provide the expected results?

A system is more than just the processes. It is the sum of processes, responsibilities, authority, procedures, and resources. Each of the elements of a system needs to be compatible and the elements coordinated. The ISO 9000 series recommends that each process have an owner as the person responsible. Each system should also have an owner as the coordinator. Strategic planning is especially important to avoiding system problems.

Some of the guidelines on quality planning given by the ISO 9000 family should be considered in planning the environmental management system as well:

- preparation of plans.
- identification and acquisition of controls, processes, equipment, fixtures, resources, and skills needed (including inspection and test equipment).

- ensuring the compatibility of procedures for design, the production process, installation, servicing, inspection, and applicable documentation.
- updating or developing techniques when needed.
- identifying measurement requirements that exceed the known state of the art in time for the needed capability to be obtained or developed.
- identifying suitable verification points in the production and processing of product.
- clarifying standards of acceptability for features and requirements[10]
- identifying and preparing records.

Establishing an efficient way to identify environmental aspects and evaluate impacts is imperative because "We are moving toward global ecological reform. It will be global, and it will be startling. We're coming to the precipice, and we're going to come to the conclusion that we want to stick around" (Hawken 1993).

We are each responsible for the management of our impacts on the environment. "The word 'crisis' is over-used in the rhetorics of environmental concern. However, thinking about the 'crisis' in its twin dimensions of *danger* and *opportunity* can help us to better locate and understand both the jeopardy we all face in the current political time and how our own actions can continue to make a critical difference" (Morgan-Hubbard 1995).

4.2.2 Handling the Environmental Laws

It is recommended by ISO 14000 and required by ISO 14001 that organizations establish and maintain procedures to identify, have access to and understand all legal and other requirements "to which it subscribes, directly applicable to the environmental aspects of its activities, products, and services."

To maintain regulatory compliance you need to identify and understand requirements applying to your activities, products, and services. Regulations usually take one of these forms:

- those specific to an activity, like a site operating permit,
- those specific to product or service,
- those specific to your industry,
- general environmental laws, or
- authorizations, licenses or permits.

As for sources to contact to identify environmental regulations and changes to regulations, you can consider any level of government, industry associations, commercial databases, or professional services. Every organization should establish and maintain a list of applicable laws and regulations to track changes.

British Standard 7750 requires you have a procedure and maintain a record of applicable requirements. It also requires that contractors be kept aware of relevant requirements. Legislative and regulatory requirements may apply to planning conditions, consent to discharges, process authorizations, improvement notices, and other vehicles. By maintaining the register of applicable requirements, the organization has hard evidence it is aware of its environmental obligations.

Contracts must be reviewed by an organization before they are accepted, according to ISO 9000. This review will ensure the requirements are defined and agreed to, differences are resolved, and that the organization is capable of meeting the requirements, with consideration of the economics and the risks.

The implications of ISO 14000 for U.S. power generators are broad, according to Jason Makansi. His Legislative Update column in *Power* magazine discusses the implications of ISO 14000 when considered together with current legislative trends nationwide and criminal sentencing guidelines for noncompliance with environmental statutes.

> Simply, ISO 14000 formalizes the development and implementation of corporate policies and procedures for environmental management and auditing practices to ensure conformity.... Implicit in the ISO 14000 exercise is that voluntary disclosure and compliance will ultimately become a competitive advantage (Makansi 1995).

According to Makansi, no one expects ISO 14000 to replace U.S. regulations for specific pollutants or discharges, but the Environmental Protection Agency and some U.S. states see it as "due diligence" (Ibid.). Compliance with ISO 14000 "is a potential indicator that a company has achieved a certifiable level of environmental leadership." It is "a means of showing that a company is making good-faith efforts to meet environmental requirements. Others might see it as 'posting bail on the lay-away plan'." But for the short-term, it could help keep people out of jail.

The federal sentencing guidelines state:

> An effective program to prevent and detect violations of law means a program that has been reasonably designed, implemented, and enforced...failure to prevent or detect the instant offense, by itself, does not mean that the program was not effective. The hallmark of an effective program...is that the organization exercised due diligence in seeking to prevent and detect criminal conduct by its employees and other agents (Ibid.).

According to James Levin of A. T. Kearney in Alexandria, Virginia, the ISO 14000 standard is "a potential source of competitive advantage" (Ibid.). In the globalization of electric power markets, compliance with ISO 14000 will level the playing field for world trade. "Private investment in powerplants, sweeping the world, makes the importance of a standard like ISO 14000 obvious," Kearney writes (Ibid.).

It will become "a calling card a company must hold to do business as a multinational," writes Makansi. Look at how much the quality management standard, ISO 9000, permeates manufacturing today.

> Thus, ISO 14000, and countless individual and collective efforts like it, may simply reflect the fact that responsible management of the environment is a cultural trait that must become as natural as shaking hands, bowing, or embracing when doing business in the global village (Ibid.).

The risks, according to Makansi, involve lifting relief from exemptions on reporting pollutant discharges under the Toxic Release Inventory program. Some companies may lack the proper information on what their environmental problems are, and the first environmental review may be shocking if publicized. Utilities, now preoccupied with the shift to competition, may see the disclosure and sharing of information implied by ISO 14000 as a competitive threat.

Japan was represented at the 1992 Earth Summit and had "an unusually high profile, clearly indicating that the environmental wave is one that Japan would like to ride" (McInerney and White 1993). In Japan, the Keidanren, the club for major corporations issued in 1991 a Global Environmental Charter with guidelines to encourage good environmental behavior among its members (Ibid.).

I believe that at least the intent of the Rio Declaration will find its way into international laws in the near future. At least three fourths of consumers call themselves environmentalists (Ibid.).

> [Z]ero emissions requires new thinking about how matters are handled throughout the manufacturing process.... Pollution prevention doesn't always pay when measured

> in conventional terms. Looked at in isolation or evaluated by strict financial standards, waste reduction can look like a poor investment. But without a philosophy of lean management, waste-free production will *always* produce results (Ibid.).

J. Gordon Arbuckle and Thomas F.P. Sullivan report on environmental law as a system (Arbuckle and Sullivan 1991). They state that environmental law has evolved into a "*system* of statutes, regulations, guidelines, factual conclusions, and case specific interpretations which relate one to another in the context of generally accepted principles established during the history (short though it may be) of the field" (Ibid.). Understanding this complex environmental law system is a definite challenge. The system "is an organized way to minimize, prevent, punish or remedy the consequences of actions which damage or threaten the environment, public health and safety" (Ibid.).

According to Arbuckle and Sullivan, there are eight generic compliance obligations—regulatory approaches—utilized (Ibid.):

1. Notification requirements—to advise authorities, employees and the public of intended or actual environmental impacts.

2. Point of discharge controls—to prevent or minimize environmental impact.

3. Process-oriented controls (includes pollution prevention)—to reduce quantities, prevent releases, and minimize hazardous characteristics of wastes generated.

4. Product-oriented controls—to assure products are designed, formulated, packaged, or used with no unreasonable risk to the environment or human health.

5. Regulation of activities—to protect resources, species or ecological amenities.

6. Safe transportation requirements—to minimize risks inherent in transporting hazardous wastes or materials or other products.

7. Response and remediation requirements—to clean up released pollutants, prevent threat of release, or pay for clean up or prevention.

8. Compensation requirements—to make those responsible pay for damages done to the environment or health, or to allow self-appointed representatives of society to recover for undue injuries.

Arbuckle and Sullivan offer some very good guidance on handling environmental laws (Ibid.).

"The fact is that many of the questions which are most critical to successful compliance efforts and most difficult for environmental practitioners to answer fall within this category:

- What level of government has authority to regulate?
- What institution [branch] of government has authority to regulate?
- What protections are available to the regulated?
- Do organizations and individuals have substantive rights?
- How do questions of scientific fact get answered?
- Who can go to court and who pays for it?"

They remind the environmental practitioner,

> The field of environmental law may involve a higher degree of incertitude than most other areas because of its newness and changeability.... Police power is the inherent right of a government to pass laws for the protection of

> the health, welfare, morals, and property of the people within its jurisdiction. Police power may not be bartered away by contract. It extends to all public needs (Ibid.).

With regard to whether your own reports (like audits and reviews) can be used as evidence against you, Arbuckle and Sullivan say, "The extent to which the results of an investigation or inspection are available in private liability litigation remains uncertain" (Ibid.). For U.S. companies they say, "A corporation is not protected by the self-incrimination provisions of the Fifth Amendment to the U.S. Constitution. So, it may not object to the production of its books to be used as evidence against it" (Ibid.).

Some final words about environmental law:

> Aggressive compliance is the most effective protection against aggressive enforcement and other efforts to assess liability.... Everyone is responsible for environmental law compliance and, to protect against individual liability, should continually demonstrate due concern and diligent efforts to comply. Providing appropriate education and training plus sufficient informational resources is a good demonstration of concern for compliance.... The best answer to the question of what can be done to prevent violations and minimize liability is an appropriate corporate 'culture' or management structure formulated with a view to environmental objectives and aggressively implemented.... After your compliance system is in place, periodic 'audits' to verify compliance and identify areas where compliance can be improved will be helpful (Ibid.).

4.2.3 Establishing Performance Criteria

In the absence of external environmental standards meeting the needs of an organization, internal priorities and performance criteria should be developed and implemented. Then, together with existing external standards, the performance requirements for the organization can be defined to meet policy and environmental objectives and targets. Here are typical areas where an organization may have internal performance criteria, according to ISO 14000:

- management systems
- employee responsibilities
- acquisitions
- property management
- divestitures
- suppliers and contractors
- product stewardship
- regulatory relationships
- process risk reduction
- environmental communications
- pollution prevention
- resource conservation
- incident response
- incident preparedness
- awareness and training
- capital projects
- process changes
- waste management
- hazardous materials management
- environmental management and improvement
- water quality management
- air quality management
- energy
- transportation management

It is important to consider performance criteria to measure the effectiveness of key personnel. British Standard 7750 assigns certain responsibilities to management and other personnel, and their performance criteria would include roles like these:

- Senior management is responsible for developing, resourcing, reviewing, and complying with the environmental policy.
- The management representative is responsible for ensuring that changes in environmental legislation and regulations

are monitored, evaluated and brought into the environmental management system.

- Personnel are responsible for developing and maintaining two-way communication and training programs on matters of environmental management.

- Accounting procedures are to be developed and maintained to identify costs and benefits relating to environmental management.

Carol Kennedy, the executive editor of *Director*, the journal of the Institute of Directors in London, has a book on the "best ideas" from people who have changed the way we manage. "There is a limit to the number of original ideas in any field of human activity, and management is no exception" (Kennedy 1991). Key management experts cited below have "best ideas" that apply to designing internal performance criteria.

John Adair, the creator of the Action-Centered Learning model, identifies six elements of note (Adair 1990):

1. Planning
2. Initiating
3. Controlling
4. Supporting
5. Informing
6. Evaluating

Chris Argyris, organizational psychologist, states "The significant human relationships are the ones which have to do with achieving the organization's objective" (Kennedy 1991). Performance depends on communication of the criteria and the evaluation measurements. Chester Barnard notes the importance of communications (Ibid.):

- everyone should know what the channels of communication are,

- everyone should have access to a formal channel of communication, and

- lines of communication should be as short and direct as possible.

Warren Bennis, former Presidential adviser, believes "Managers do things right. Leaders do the right thing," and identifies the leader as the "social architect" (Ibid.).

There are seven essential skills for the performance of managers of the future, according to American sociologist Rosabeth Moss Kanter (Ibid.):

- Learn to operate without the hierarchy "crutch."

- Know how to compete in a way that enhances, not undercuts, cooperation.

- Operate to the highest ethical standards.

- Possess a dose of humanity.

- Develop a process focus on how things are done.

- Be multifaceted and ambidextrous; work across functions to build synergies.

- Be able to gain satisfaction from results and be willing to stake your own rewards on them.

Robert Sibson, founder and CEO of Sibson and Company, one of the world's largest human resources consulting firms, has his own ideas.[11] He says, "Ideally, a commitment to work excellence by executive management must be so clear and persuasive that a desire for excellence exists throughout the organization. In my mind, this is the essence of organizational leadership.... A commitment to excellence includes a commitment to hard work. That doesn't mean a sweatshop or even a stressful environment. But you will never find high levels of effectiveness by any person or group where there is not a commitment to hard work" (Ibid.).

Other advice from Sibson to be considered in designing your internal performance criteria hinge on an understanding of the "effectiveness ethic." He writes, "The effectiveness ethic is a set of views, attitudes, and values that induces people to do their best at work. The effectiveness ethic is an inclination to do your best each and every time. When there is an effectiveness ethic, those attitudes are as automatic as breathing. They become so much a part of the consciousness of employees that doing one's best is the natural way and the only way" (Ibid.).

Your organization's people are your success or your failure. The days of robotic human performance are over. As every one of us is given more freedom to access others and compare our fate with theirs, there is no question we will all demand more.

Sustainable development has to do with improving everybody's quality of life, something that is an evolutionary right and purpose by universal design. Your environmental performance is intimately tied to your people's performance.

Sibson indicates that the effectiveness ethic is different than the work ethic (Ibid.). It focuses on rewards for work excellence and the need for good work rather than on an obligation to work. "When there is an effectiveness ethic, workers constantly seek a better way to accomplish required work" (Ibid.). Sibson believes if you seek an effectiveness ethic, you're more likely to get one if you show a commitment to something worthwhile. In his opinion, "High goals can cause excitement.... High but attainable goals are motivational" (Ibid.).

According to Sibson, "leaders at the operating level always possessed the following two characteristics:

1. They managed successful operations, and at least part of the success was because of higher organization productivity.

2. Workers think the managers made them better and more effective, and encouraged them to strive for excellence" (Ibid.).

Sibson warns managers to attack unproductive or obsolete practices. He recommends deregulating the organization as well, "Bureaucracy is its own form of an unproductive practice" (Ibid.). Sibson suggests using knowledge "to gain greater effectiveness of work" and says it involves four elements:

- Getting knowledge,

- Managing knowledge,

- Managing knowledgeable workers, and

- Networking, including the use of experts.

If you can spot the winners in your organization, you need to also spot the losers. What is the difference between them? Unless you can identify the difference and measure it, maybe there is no difference. Or maybe there is no competition, and the notion of winner/loser is one you should drop. The team concept is the current vogue and whether we like it or not, the advance of society into the coming age of incredible scientific and technological development will mandate team work. If you and other organizations around the world create the standards you live by, then governments will have less justification to set them. Effective environmental management which includes solid performance criteria will help achieve this goal.

4.2.4 From Policy to Objectives to Targets

The policy will generate environmental performance goals and objectives that the organization sets out to achieve. The results of the initial environmental review need to be taken into consideration as well as any environmental aspects or impacts identified. Measurable performance indicators need to be developed as a basis for the objectives and to monitor performance against the objectives.

Objectives, according to ISO 14000, include commitments such as the following:

- Reduce waste;
- Reduce resource depletion;
- Reduce or eliminate environmental pollution;
- Design products for minimal environmental impact in production, use and disposal;
- Control environmental impact of raw material sourcing;
- Control environmental impact of new developments;
- Promote environmental awareness among employees; and
- Promote environmental awareness within the community.

Notice that those examples are distinct but somewhat generic in scope and purpose. Each needs to be further defined; for example, the first may be a policy statement that would be refined into a specific objective, such as to reduce waste wherever technologically or commercially viable and to reduce resource depletion during production activities.

Environmental targets should be set for the identified objectives within a specified time frame. Targets need to be specific and measurable. Both the objectives and the targets can apply across the organization or to narrower, site-specific or individual activities. Appropriate management at all levels should have the responsibility for defining environmental objectives and targets. And they need to review them periodically for any needed revisions. Always consider the views and concerns of interested parties.

Here are suggested environmental performance indicators to measure and observe trends toward achieving objectives, i.e., environmental targets:

? quantity of raw material used

? quantity of energy used

? quantity of emissions

? waste per quantity of finished product

? efficiency of material and energy use

? number of environmental incidents/accidents

? number of prosecutions

? number of vehicle miles per unit of production

? specific pollutant concentrations

? land area for wildlife habitat

? percent of waste recycled

? percent of recycled material used in packaging

? investment in environmental protection

An environmental objective becomes a performance indicator, which becomes an environmental target. An example given in ISO 14000 is shown below.

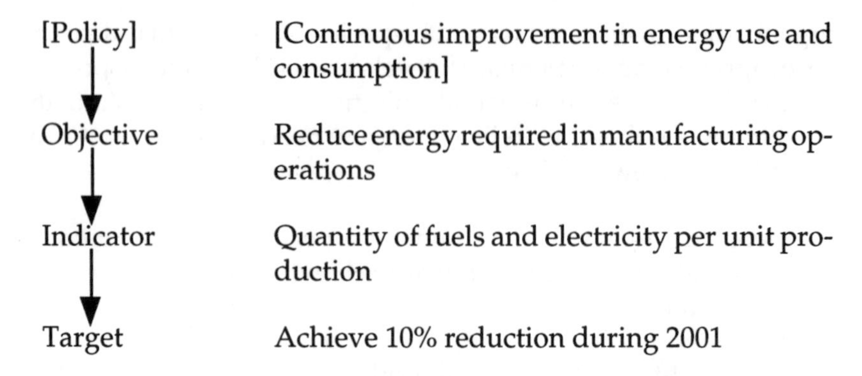

[Policy]
[Continuous improvement in energy use and consumption]
Objective
Reduce energy required in manufacturing operations
Indicator
Quantity of fuels and electricity per unit production
Target
Achieve 10% reduction during 2001

In developing environmental objectives, BS 7750 says that environmental impacts need to be evaluated and codes of practice should be examined. You may have committed to achieving certain levels of control over adverse impacts or to specifying numerical targets for pollutant load reduction or waste reduction. You may have committed to installing and using certain types of technology. It is also recommended by BS 7750 that the views of interested parties be considered, along with the frequency and nature of any complaints about your environmental performance.

Both direct and indirect environmental aspects need to be considered. Direct aspects may involve disposal or release of wastes from production processes (liquid, gaseous, or solid); use of resources (fuels, energy, and materials); effects of transportation; and the effects of land management practices. Indirect aspects may involve the extraction of raw materials supplied by another organization, the effects of other businesses where your reserves are invested, or the aspects of use (and possible misuse) of products, i.e., all those environmental aspects and associated environmental impacts that you control and have influence over.

The section on environmental effects is one of the longest in BS 7750. Guidance is provided on considering the potential impacts of suppliers, the ghosts of the past (liability for former products), and the consequences of activities transferred through acquisitions as well as future consequences that may arise from present activities. It is imperative that the severity of potential environmental impacts be considered as a basis for establishing environmental objectives. The severity will influence the control of risks.

Pertinent questions to ask when evaluating environmental impacts include:

? What is the likely significance of the impact on the environment? On organisms? On the ecosystem?

? What are the regulatory requirements?

? What are the concerns of interested parties?

? What is already known about the environmental impact?

? Has the risk been assessed?

? Has the environmental impact been fully identified and evaluated?

The evaluation of environmental aspects and environmental impacts, both real and potential, allows the organization to identify environmental objectives and targets. Areas targeted for improvement should be those with the most need to reduce risks and liabilities. A cost-benefit analysis is recommended by BS 7750.

Objectives and targets should be quantifiable. Environmental objectives should be set to realizing improvements in performance over time based on economics, available technology, the degree of environmental impact, and the risks involved. Environmental targets need to be demanding and achievable.

Your environmental management policy will set your objectives and targets and demonstrate to everyone inside and outside your organization the direction you're headed. Live up to your policy to keep a competitive advantage.

4.2.5 From Environmental Targets to Actions

Midway through ISO 14000 is a section on Environmental Management Plans and Programs. Organizations are encouraged to establish environmental management plans and programs to address schedules, resources and responsibilities for achieving the environmental policy, objectives and targets. The environmental management plan provides a long-term framework for improving environmental performance. The plan may be free standing, although integrating the plan into the organization's strategic plan would have greater benefits.

The environmental management program inside the environmental management plan identifies the specific action steps, schedules, and resources and responsibilities required to make the targets. It does so in the order of their priority, set by management.

Environmental management plans and programs need to be dynamic and revised regularly to reflect changes in environmental objectives and targets. Remember, the plan is long term and the program is short term. The long-term planning helps in mapping out the process for continual improvement in environmental performance. An environmental management plan may include these items:

- Description of current environmental performance.
- Description of proposed performance improvements planned.
- Consideration of technology requirements needed to implement proposed improvements.
- Awareness of internal and external factors influencing proposed improvements.
    - ♦ financial resources
    - ♦ capital projects
    - ♦ facility location
    - ♦ market needs/developments
    - ♦ legislative/regulatory developments
    - ♦ expectations of interested parties

The environmental management program, with short-term goals, may include items such as

- prioritization of environmental issues,
- development of options for priority issue resolution,
- cost/benefit review of options,
- selection of preferred approach from options,
- identification of responsibility for implementation, and
- provision for post-implementation review and assessment.

An example given in ISO 14000 of how to link environmental objectives and targets to environmental management plans and programs is shown below.

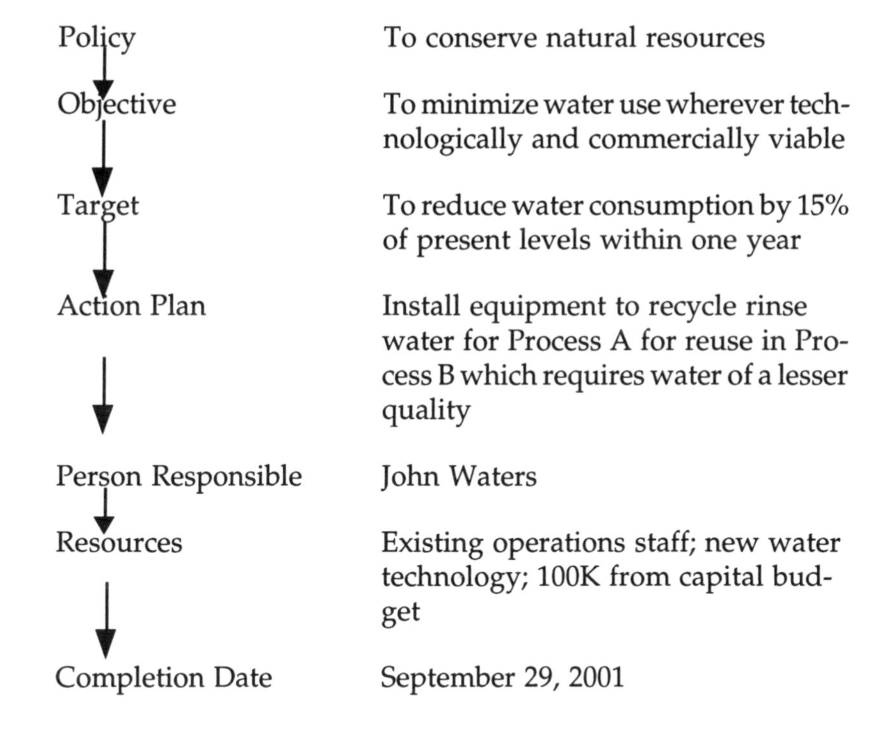

Policy
To conserve natural resources
Objective
To minimize water use wherever technologically and commercially viable
Target
To reduce water consumption by 15% of present levels within one year
Action Plan
Install equipment to recycle rinse water for Process A for reuse in Process B which requires water of a lesser quality
Person Responsible
John Waters
Resources
Existing operations staff; new water technology; 100K from capital budget
Completion Date
September 29, 2001

Similar to the requirements of ISO 9000, ISO 14001 requires top management to appoint a specific management representative, who has a defined role, responsibilities, and the authority for ensuring requirements are established, implemented and maintained, and the obligation to report on environmental performance for top management review.

In BS 7750, there is a requirement for a management representative with defined authority and responsibility to ensure the requirements of the standard are implemented and maintained. This individual is to have sufficient knowledge of activities and environmental issues to be effective. This individual should coordinate implementation of the requirements of the standard, with line management fully responsible for implementing the environmental management system.

Staying on top of additions and changes to environmental legislation and regulations is a primary responsibility of the management representative. The management representative should also keep up on the concerns of interested parties. Every stakeholder deserves attention. Every stakeholder deserves successful implementation of your policy and plans for environmental management. You are the mover, the shaker, and the motivator.

4.3 Implementing an Environmental Management System

Principle 3 introduces Section 4.3 of ISO 14000:

> **For effective implementation an organization should develop the capabilities and support mechanisms necessary to achieve its environmental policy, objectives, and targets.**

This section discusses the resources and support mechanisms needed to ensure the capability of the environmental management system:

Ensuring Capability

- Resources—Human, Physical and Financial
- System Alignment and Integration
- Accountability and Responsibility
- Environmental Values and Motivation
- Knowledge, Skills and Training
- Support Action
    - Communication and Reporting
    - Documentation
    - Records and Information Management
    - Operational Controls
    - Emergency Preparedness and Response

Consider this simple analogy. You want to dig a hole in your backyard to plant a tree. There's a shovel in your garage. Unless you bring the shovel to the backyard and shove its blade into the ground, there will be no hole. No hole, no tree. To implement means to use. To implement means to manage. Manage, or the result is chance and chaos.

4.3.1 Ensuring Environmental Management Capability

Because of changes in the requirements of interested parties, the dynamic business environment, and the process of continual improvement, an organization's capabilities and support mechanisms must constantly evolve in response. An organization needs to focus and align its people, systems, strategy, resources, and structure.

Implementing environmental management is best approached in stages, based on the level of awareness of requirements, environmental aspects, expectations, benefits, and the availability of resources.

Ensuring the environmental management capability of an organization involves a concentrated effort on the part of upper and middle management. Every employee should provide input to define the necessary resources and allocate them effectively. Training needs for each person in the organization involved in environmental management (and that's just about everybody) should be identified. Minimum capability expectations should be included in your policy and objectives.

It may be that you will want to outsource certain environmental management functions to obtain the utmost capability from experts. For example, you may want to procure third-party certification that environmentally labelled products meet a set of predetermined criteria. Green Seal is an independent group in the United States that awards a green-colored seal of approval to products. These products cause far less impact on the environment than other similar products. Or you may want self-declaration labelling, such as "biodegradable" or "environmentally friendly" (*Export Today* 1995).

The following sections on resources, system alignment and integration, accountability and responsibility, values and motivation, and knowledge, skills and training, are important to ensuring your capability as an environmental manager. Reducing the risks to your organization and those *created* by your organization is your fundamental management task.

Dr. Allan Cohen considers vision an important part of implementing your management system. Vision is the "enrichment of the mission with energy and color" (Cohen 1994). The vision for the original AT&T, "We will build a telephone system so that anyone, anywhere in the world, can talk with anyone else, cheaply, quickly, and satisfactorily" (Ibid.), is an excellent example of organizational vision.

> Empowerment exists in an organization when lower level employees feel that they are *expected* to exercise in good faith on behalf of the mission even if it goes outside the bounds of their normal responsibilities; and if their initiative should lead to a mistake—even a serious one—they trust that they will not be arbitrarily penalized for having taken the initiative.... When you see something that needs to be done, do it! Don't wait to be told to do it, don't sweep it under the rug, don't blame it on someone else (Vaill 1993).

Dr. Todd Jick proposes ten "commandments of managing change" (Jick 1993):

1. Analyze the organization and its need for change.
2. Create a shared vision and common direction.
3. Separate from the past.
4. Create a sense of urgency.
5. Develop a strong leader role.
6. Line up political sponsorship.
7. Craft an implementation plan.
8. Develop enabling structures and reinforcements.
9. Communicate, involve people, and be honest.
10. Monitor, refine, and institutionalize change.

Dr. Rosabeth Moss Kanter, offers some keen guidance for environmental managers on the economics of the future.

> Globalization brings managers two almost contradictory challenges. It increases the need for cooperation and coordination among businesses and among countries in order

> to find common standards, methods, languages, package sizes, transportation systems, or communications links. Indeed, to get access to all the cash, skills, goods, or markets the world can offer, globalization encourages companies to form relationships across traditional boundaries—alliances that can bring additional benefits quickly.... Skills in influence, negotiation, and conflict management are not optional for global managers...they are necessities.... Managers have to think internationally even if still operating just a local department in a local company—not because their *company* has locations elsewhere but because their *partners* do (Kanter 1993).

Remember, management means reducing the risks to your competitiveness, to the environment, to your stakeholders, and to yourself.

4.3.2 Knowing What Resources Are Needed

Appropriate human, physical, and financial resources that are essential to implementing and achieving environmental objectives need to be defined and made available. It is easier to accomplish this if procedures are first developed for allocating resources to track the benefits as well as the costs of pollution control, wastes and disposal. According to ISO 14000 the following entities need to be considered in order to manage constraints against resource allocations:

- Large client organizations
  - to share technology
  - to share know-how
- Other organizations in your supply chain or local organization
  - to define and address common issues
  - to share know-how

  - ◆ to facilitate technical development
  - ◆ to use facilities jointly
  - ◆ to study the environmental management system
  - ◆ to collectively engage consultants

- ➥ Standardization organizations
- ➥ Industry associations
- ➥ Chambers of Commerce
- ➥ Universities and research centers for support of production and innovation

Take an inventory of what resources are needed for effective environmental management. It could be that fewer are required than you think if the resources are already available. In any case, look first at those resources that would give you the most sophisticated system ever, then at those needed to meet the minimum requirements and expectations. Shoot for the highest, most conservative compromise you can. To know what is enough, you need to know what is more than enough.

Be as generous as you can with your environmental management. The ecosystem and the Earth's ecology can probably do without us as a species. We are expendable. Protect the environment—it's self-defense and common sense.

4.3.3 Aligning and Integrating Management Systems

Elements of the environmental management system should be designed or, if necessary, revised so that they are effectively aligned and integrated with existing management elements. These are the elements which may require integration, according to ISO 14000:

- ★ Policies
- ★ Resource allocation
- ★ Operational controls
- ★ Training
- ★ Organizational structure
- ★ Appraisal systems
- ★ Measuring systems
- ★ Communications
- ★ Documentation
- ★ Information systems
- ★ Support systems
- ★ Employee development
- ★ Accountability structure
- ★ Rewards systems
- ★ Monitoring systems
- ★ Reporting

Merge your management systems and controls so they feed each other. Each system should assimilate important points from other systems where they intersect or share resources. Take a close look at the flow of each management system within your organization and the main flow of all of the systems and the synergies between them.

4.3.4 Assigning Responsibilities and Accountability

As mentioned early in this primer, responsibility for the overall effectiveness of the environmental management system needs to be assigned to the most senior person with sufficient authority, competence and resources. Responsibilities of personnel who implement the system should be clearly defined by operational managers who are held accountable for effective environmental performance. Every employee with a responsibility for environmental performance must also be held accountable within the scope of their involvement.

It is necessary, according to ISO 14000, to assign appropriate responsibilities and authority to ensure an effective environmental management system is developed and implemented. The Practical Help of Section 4.3.2.3 suggests a possible approach, recognizing that companies and institutions have different organizational structures, with different needs in understanding and defining environmental responsibilities for their own work processes. Here's how ISO 14000 assigns environmental management roles and responsibilities.

Environmental Responsibility	Person Responsible
Establish overall direction	President, CEO, Board of Directors[12]
Develop environmental policy	President, CEO, Chief Environmental Manager[13]
Monitor overall system performance	Chief Environmental Manager, Environmental Committee[14]
Assure regulatory compliance (external)	Senior Operating Manager
Ensure internal compliance	All Managers, Chief Environmental Manager[15]
Ensure continual improvement	All managers[16]
Identify customers' expectations	Sales and Marketing Staff
Identify suppliers' expectations	Purchasers, Buyers
Develop and maintain accounting procedures	Finance/Accounting, Managers
Comply with defined procedures	All Staff

In the case of small or medium-sized enterprises, it may be the owner who has the responsibilities and is held accountable for these activities. In any case, you as the most senior manager, as *leader*, are ultimately accountable for your organization's impacts on the environment, both good and bad. When you delegate responsibilities, be sure to delegate sufficient authority for making decisions and taking necessary actions. Success is guaranteed only by how well you delegate management.

4.3.5 Encouraging Environmental Management Values

It is recommended by ISO 14000 that a common set of environmental values be developed and reinforced, taking into account the views of interested parties. Senior management plays a key role in communicating values and in motivating employees to perform.

But it is the commitment of individual people that transforms the environmental management system from just paperwork into an effective process, especially in the context of shared values.

All members of the organization should understand and be encouraged to accept accountability for achieving environmental objectives and targets. In turn, each individual should encourage others to respond in a similar manner. The best motivation for continual improvement is through recognition and reward for achievement and for making suggestions that can lead to improved environmental performance.

"Values enter into practically every decision a manager makes." A value system is "an enduring organization of beliefs concerning preferable modes of conduct or end-states of existence along a continuum of relative importance." "Terminal Values" are the ends toward which you strive, and "Instructional Values" are the means to achieve the ends (Hitt 1990).

"Leadership is seductive. Management is work" (Sutton 1993). Motivate, first; then you can manage. Your values will become the values of everyone in your organization if you motivate correctly.

4.3.6 Assuring Know-How and Competence

The knowledge and skills needed to achieve environmental objectives must be identified. Required competence is important to know when selecting and recruiting personnel, determining training to be offered and encouraging continued education in certain subjects.

Employees need to have an appropriate knowledge base in order to achieve environmental objectives and targets. This knowledge base includes training in methods and skills required to perform their tasks efficiently and competently. It also includes knowledge of the impact of their activities on the environment, especially if performed incorrectly. The knowledge base of responsible contractors should also be addressed.

The level and detail of training varies according to the task, of course, so it is essential that the employee's knowledge of regulatory requirements, internal standards, policy, objectives, and targets be established, reinforced and maintained appropriately. Here's what ISO 14000 recommends as elements within effective training programs:

- Identification of each employee's training needs.
- Development of a training plan to meet needs.
- Verification of compliance of training program to regulatory or organizational requirements.
- Training of target employee groups.
- Documentation of training received.

There are five basic types of environmental training that should be provided as a minimum, according to ISO 14000. The training, intended audience, and purpose are shown below.

Type of Training	Audience	Purpose
Raising awareness of the strategic importance of environmental management	Senior Management	To gain commitment and alignment to the organization's environmental policy
Raising general environmental awareness	All	To gain commitment to the environmental policy, objectives and targets and to instill a sense of individual responsibility
Skills enhancement	Employees with environmental responsibilities	Improve performance in specific areas—operations, Rand, engineering
Compliance	Employees whose actions can affect compliance	Ensure regulatory and internal requirements for training are met

The competence of personnel performing tasks that may cause significant environmental impacts is important in the eyes of ISO 14001 and the world. To ensure the appropriate amount of training is provided, the specification recommends employee awareness of the following:

- Importance of conformance with environmental policy.
- Importance of conformance with environmental management requirements.
- Actual and potential significant environmental impacts associated with work.
- Benefits of improved personal performance.
- Roles and responsibilities, including emergency preparedness and response.
- Consequences of departure from standard operating procedures.

British Standard 7750 wants procedures in place to ensure that employees and members of the organization at every level are aware of those same issues. Training needs must be identified and training provided, with records of the training maintained. Competence comes from education, training and experience. Liability of management comes from failure to provide required training.

Executive and management personnel may need training on the environmental management system so they have the knowledge to play their part and understand criteria for judging effectiveness. Other personnel may need training on the system so they can make meaningful contributions to its effectiveness. New employees will need training on new tasks, unfamiliar equipment, procedures. In all cases, employees should be motivated to having the proper regard for environmental concerns and be encouraged to participate in environmental initiatives.

Clay Carr offers the following specific ideas on competence (Carr 1992). There are five categories of competence and each is important to the core competence of an organization:

- job competence
- interpersonal and communication competence
- background competence
- organizational competence
- self-management competence

Carr thinks, "The core competence that your firm develops must derive from its deepest strategic understanding of itself and its market." There is a competitive advantage of understanding your core competence. "The ability of any firm to compete in today's dynamic economy stems from its ability to develop, maintain, and increase the competence necessary not only to preserve its current position, but to identify and respond successfully to the new opportunities the market presents" (Ibid.).

But your core competence is no static entity.

> The usable competence of your organization never stands still; it's constantly increasing or decreasing in competition with your competitors and your market. When competence is supported by confidence and given opportunity, your firm's mastery spirals upward. When competence, confidence, or opportunity is missing, the spiral heads toward the basement (Ibid.).

> Effective training specifically supports the organization's improvement strategy.... Training is often an important form of communication and resource sharing.... Supporting a compelling vision with a set of clear goals provides the direction people need to understand where the company is trying to go (Cocheu 1993).

People are members of your organization for any number of reasons. To earn a wage is definitely the main reason for most, probably even your own. But there's more to a job than the paycheck. Some people see their work as simply a job, and others see it as their career. In both cases, maintaining skills and developing new skills adds value, both to the individual's capabilities and to your resource base. Cross-functional training expands your workforce without expanding staff. Most of the talents of each individual within your organization are a unique part of the culture of your organization. And, like history, most of the talents must be passed down and passed on to younger generations in order to be maintained. So make sure everyone who performs in your environmental management system wants to come to work for more reasons than just the money. Adopt an organization-wide philosophy: I'd love working here for a lifetime.

4.3.7 Supporting Your Environmental Management System

Organizations need to identify and plan those functions, activities and processes that do or may have significant environmental impacts. The functions and activities must be carried out under controlled conditions with particular attention to documented procedures and instructions, process monitoring and control, and performance to specified criteria. Compliance with requirements needs to be communicated to responsible individuals and their performance verified. Supporting the environmental management system will involve at least the following actions:

- Identify and document the information to be obtained and verified; specify the required accuracy of the results.
- Specify and document verification procedures; identify the locations and times of measurement.
- Document and maintain quality control procedures, calibration and control charts, and maintain records.

- Document procedures for data handling and interpretation.

- Document acceptance criteria and actions to be taken for unsatisfactory results.

- Assess and document the validity of data from malfunctioning verification systems.

- Safeguard measurement and test processes from unauthorized adjustment or damage.

It should always be the objective to control the activity in question in accordance with specified requirements and to verify the outcome. When there may be indirect environmental impacts of an activity, the control and verification procedures should address those functions, activities, and processes by which the organization can exert influence. For example, BS 7750 suggests that if your policy is to provide customers with information on the environmentally responsible use of your products, procedures should control and verify providing the information. If your policy is to buy materials from companies with sound environmental performance, procedures should obtain performance information from the suppliers. In any case, what you want is to be able to demonstrate adherence to policy requirements.

The concept of sustainable development increases your family. Your organization is your family. Your customers are your family. Your stakeholders are your family. Support them all as carefully and fully as you do your own children. Try it and see how much you motivate repeat sales, new customers, improved productivity, and improved environmental performance. Never forget, everyone in your organization depends on your support as much as your leadership in order to support you. Share yourself with them, and you will understand each of them better. Listen, and your voice is heard.

4.3.8 Handling Communications

According to ISO 14000, communication includes establishing processes and plans to report internally and externally on environmental activities of the organization for the following reasons:

- to demonstrate management commitment to the environment.
- to deal with concerns and questions about environmental issues.
- to raise awareness of environmental policies, objectives, targets, and programs.
- to inform interested parties about the organization's environmental management system and environmental performance.

Information to be communicated includes results of monitoring system performance, audits and management reviews. Appropriate information should also be provided to employees to motivate them and to external interested parties to encourage public understanding and acceptance of the organization's efforts to improve environmental performance.

An organization can communicate environmental information in a variety of ways, according to ISO 14000:

- Externally through:
    - Annual reports
    - Regulatory submittals
    - Public records
    - Industry association publications

  - ♦ Media
  - ♦ Paid advertising[17]

- Open house
- Published telephone/FAX numbers for complaints/questions
- Internally through:
  - ♦ Bulletin board postings
  - ♦ Newsletters
  - ♦ Meetings
  - ♦ Electronic mail messages (E-mail)

Two-way communication, internally and externally, is encouraged by ISO 14000. Any information communicated needs to be understandable and adequately explained. Information should be consistent, and verifiable, and should present an accurate picture of environmental performance. The standard provides a few examples of the type of information that may be included in reports:

- Organization's profile
- Environmental policy
- Environmental objectives
- Environmental targets
- Environmental management processes
  - ♦ Interested party involvement
  - ♦ Employee recognition

- Environmental performance evaluation
  - Releases
  - Resource conservation
  - Compliance
  - Product stewardship
  - Risk
- Opportunities for improvement
- Independent verification of the contents

To borrow a phrase from a friend, we only occasionally talk the truth we have in mind.

4.3.9 Handling Documentation

Regarding environmental management system documentation, ISO 14000 recommends that operational processes and procedures be defined, documented, and updated when necessary. The various types of documents that establish and specify effective operational procedures and control should be clearly defined. It is through documentation that employees are kept aware of what's required to achieve the environmental objectives. Documentation also enables the organization to evaluate the environmental management system and environmental performance.

Where elements of the environmental management system can be integrated into documentation of the overall management system, they should. Otherwise, an "Environmental Management Manual" is recommended. For ease of use, ISO 14000 recommends organizing and maintaining a summary of the documentation as a permanent reference to system implementation and maintenance in the following manner:

- Collate the environmental policy, objectives and targets.
- Describe the means of achieving policy, objectives and targets.
- Identify key roles, responsibilities and procedures.
- Provide direction to related documents.
- Describe other aspects of the organization's management system.
- Demonstrate that elements of the environmental management system are appropriately implemented.

As for practical help, ISO 14000 offers this advice:

✍ Identify all documents by organization, division, function, activity, contact person.

✍ Obtain approval of documents by authorized personnel prior to use.

✍ Date all documents and revisions.

✍ Periodically review documents for necessary revision.

✍ Make current versions of documents available at locations where work is performed.

✍ Remove obsolete documents promptly from points of issue and points of use.

✍ Retain all documents for specified periods.

Documentation may be in paper or electronic form, according to ISO 14001. In either case, it is to describe the core elements of the environmental management system and their interaction, and to provide direction to related documentation.

The quality management standard, ISO 9000, also requires that any obsolete documents retained for legal purposes or to preserve knowledge be identified. Changes to documents need to be reviewed by personnel within the organization who have access to pertinent background information on which to base their review and approval.

In the ISO 9000 family, the preparation and use of documentation is a dynamic activity, adding high value. Documentation is essential for at least these critical roles:

- achieving required product quality,
- evaluating management systems,
- providing continuous improvement,
- maintaining gains from improvements, and
- providing meaningful audits and evaluations of system adequacy.

There must be an appropriate balance between the extent of documentation and the extent of skills and training to keep documentation at a reasonable level and keep it maintained at the appropriate intervals.

Procedures are required by ANSI/ASQC E4 to control the flow of documentation in both print and electronic media. Documents requiring control are those specifying requirements of the environmental management system as a flowdown of the provisions and commitments of the system. This includes system documents themselves. These documents are to be controlled from identification to development, through review and approval, through distribution for use, through revision to make sure only current and correct documents are used. When a controlled document becomes obsolete or superseded, it is to be removed as quickly as possible.

Documents requiring control include written, computerized, or pictorial information that describes, defines, specifies, reports, or certifies activities, requirements, procedures, or results. That's a broad range, for sure. So, when in doubt, control it.

Examples of documents requiring control include blueprints and other design documents. They also include environmental and quality instructions and procedures. Other examples are administrative documents governing compliance with legislative and other statutory requirements.

Any document that generates important records should be controlled so there is traceability from the record to the instructions in effect when the record was made. This way, any error in the record or a recorded nonconforming condition can be investigated. For example, your customer might reject a product you delivered because it was missing a piece. With document control, the reason for the missing piece can be traced back from the record to the controls over the fabrication activities in place at the time. And the root cause of the problem can be corrected.

Computer software and hardware/software configurations for design, analysis and process control are other documents requiring control. If you develop your own software, you should have a documented and approved development methodology. The software should be validated, verified and possibly tested depending on its importance to product quality or environmental aspects.

In the United States, and for use anywhere in the world, there is the recognized ANSI/ASME NQA-2 standard for computer software. Part 2.7 of the standard provides excellent guidance on software quality assurance.

There is a document control section in ISO 14001, providing requirements for procedures to ensure documents can be located and to ensure that only current approved documents are available for use. The specification also requires that procedures and responsibilities be established and maintained concerning the creation and modification of the various types of documents generated by an organization.

A document is

> any intellectual asset of the company.... Document management moves those assets, stores them, creates indexes for them, and makes them immediately accessible by many users at any time, any where they are needed.... Building the document management system properly will organize the work around results, not tasks (Currid 1994).

An excellent example of necessary and useful documentation is widely disseminated in the *World Resources Report* of the World Resources Institute. The report is manageu by the Resource and Environmental Information group "to provide a common, accessible information base for policymakers by presenting global data sets and providing objective analyses of global environmental trends." The report is distributed widely in both developing and developed countries. "Over 2000 copies of *WRR* are sent directly to foreign government officials and others influential in the policy process. Another 2000 copies are distributed in the United States, mainly to the U.S. Congress and the United Nations. A further 3000 are distributed by the UNDP and UNEP, co-sponsors of the report, and the World Bank purchases 5000 copies for its own distribution." Single copies of the reports are available free of charge for developing country institutions. Regular catalog and bookstore sales result in another 4000 copies (Gordon and Tunstall 1995).

As we blend in with the technological advances and the expansion of global information possibilities, never forget the purpose of information and documentation: to communicate and record evidence.

4.3.10 Handling Information and Records

Records are evidence of ongoing system operation. As a minimum, ISO 14000 recommends that the following types of information be generated, used and retained:

- Legislative and regulatory requirements
- Permits and authorizations
- Environmental aspects and impacts
- Environmental audits and reviews
- Product identification and composition
- Property data
- Monitoring data
- Nonconformance details
- Incident data
- Complaints and followup details
- Environmental training activities
- Supplier and contractor information
- Inspection and calibration data
- Maintenance activities

Effective management of records is one of the keys to successful implementation of an environmental management system. Information management involves nine basic features:

- Identification[18]
- Collection
- Indexing
- Filing

- Storage[19]
- Maintenance
- Retrieval
- Retention
- Disposition

You should limit records to the extent pertinent to their application, according to BS 7750. Records need to be kept in order and designed to allow assessment of compliance with the environmental policy and the achievement of environmental objectives and targets. Records to be compiled and retained include:

- Register of legislative and regulatory requirements
- Register of environmental aspects and impacts
- Audit reports
- Reviews
- Details of failures to comply with policy
- Records of corrective actions taken
- Details of incidents and follow up actions
- Supplier and contractor information
- Inspection reports
- Maintenance reports
- Product identification and composition data

- Monitoring data
- Environmental training records

Records management also involves indexing and control of access to prevent unauthorized changes to records, according to ISO 9000. They should be readily retrievable and stored in a suitable environment to prevent damage, deterioration or loss. For example, locked and fire-proof cabinets are required for records in the nuclear industry. Whether the organization keeps the records or its suppliers is up to the organization through contractual agreements.

The information generated during work results in records. From the records, the work activities that resulted in the information can be recreated. Effective handling of information and records improves organizations' operations and assures quality control.

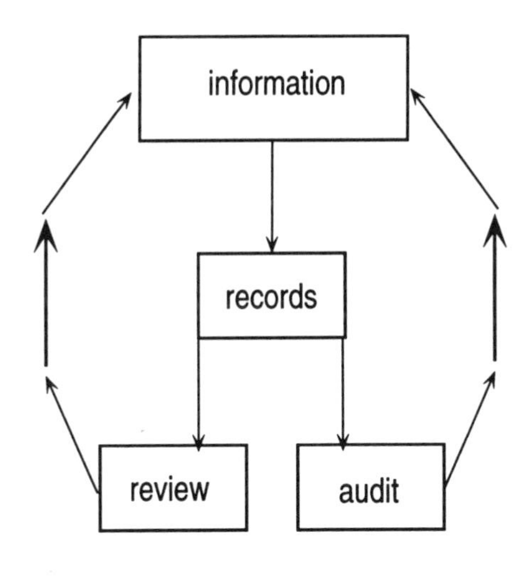

4.3.11 Controlling the Environmental Management System

Implementation of the environmental management system is through operational procedures and controls consistent with the organization's policies, objectives and targets. The various functions of an organization that contribute to significant environmental impacts need to be considered when developing procedures or controls:

- Research and development
- Design and engineering
- Purchasing and contractors
- Raw materials storage
- Production processes
- Laboratories
- Final product storage
- Transport
- Product disposal
- Customer service
- Marketing and advertising

According to ISO 14000, the activities of any organization can be divided into three categories:

1. *Activities to prevent pollution and conserve resources.* These activities apply to new capital projects, process changes, property management, new products, and packaging

2. *Daily management activities.* Management assures conformance to internal and external requirements. Management also attempts to increase efficiency and to continuously improve performance.

3. *Strategic management activities.* It is management's responsibility to anticipate and respond to changes in environmental requirements.

Statistical techniques may be appropriate, according to ISO 9000, to establish, control or verify process capability and product characteristics. Consult ISO 3534: *Statistics—Vocabulary and Symbols*, a three part standard, for guidance (Voehl, Jackson and Ashton 1994):

Part 1: Probability and general statistical terms

Part 2: Statistical quality control

Part 3: Design of experiments

If you have never used statistical process controls, I recommend you hire a retired statistician in your community to help set them up. It is cost-effective and good for both you and the retired statistician. The statistician can establish and tailor controls for your processes and product quality determinations, then train people on how to generate the control charts and how to understand them.

The basic statistical methods and applications include (Stewart, Mauch and Sttraka 1994):

- Design of experiments/regression analysis
- Pareto analysis/scatter diagrams/graphical methods
- Safety evaluation/risk analysis
- Cause-and-effect diagrams
- Quality control charts/sampling techniques
- Statistical survey techniques
- Product lifetime data analysis methods
- Time series analyses

Control comes first from the craftsperson and his or her competence. But sometimes more is required, the courage to check for substandard performance in your work. Statistics can be very enlightening when comparing actual performance to expectations.

Use statistical analysis carefully. Define and understand what is being measured or the results will be inaccurate and misleading. Don't measure the number of nonconforming items by itself; measure the apparent and true causes of the nonconformances as well. Don't measure an individual's or a process' productivity in isolation—measure the contributing causes as well.

4.3.12 Preparing and Responding to Environmental Emergencies

Emergency situations and accidents need to be anticipated with procedures and emergency plans in place to respond accordingly. These procedures and controls need to consider accidental emissions to the atmosphere, accidental discharges to water or land, and the specific environmental effects from accidental releases resulting from abnormal operating conditions. Organizations should have documented emergency response plans:

- Emergency response organization responsibilities and authority.
- Details of emergency services (fire, rescue, cleanup, etc.).
- Communications (internal and external).
- Emergency response actions (by type of incident).
- Hazardous materials information.
- Training requirements and skills determination.

The specification, ISO 14001, requires organizations to prevent and mitigate environmental impacts associated with accidents or emergency situations, and to test emergency response procedures periodically.

Being prepared for emergencies to the point of expecting them to happen, will put an organization in a better position to avoid them. When they hit, though, with a well-exercised response plan and organization, they will mitigated quickly. Each emergency response is a chance to test your skills and show your talents. Be sure to evaluate each one after you've dealt with it, and brainstorm for improvements.

4.4 Measurements and Evaluations Required for an Environmental Management System

Principle 4:

An organization should measure, monitor, and evaluate its environmental performance.

This brief section of ISO 14000 covers:

- **Measuring and Monitoring Ongoing Performance**
- **Audits of the Environmental Management System**

Measuring, monitoring and evaluating performance are essential to understanding whether an organization is complying with imposed requirements and if it is successfully pursuing its environmental objectives and attaining environmental targets.

4.4.1 Measuring and Monitoring Ongoing Environmental Management Performance

The environmental management system needs elements for measuring and monitoring actual performance and effectiveness. Then the results should be analyzed to identify needs for preventive or corrective action for improvement. There need to be processes that ensure the reliability of collected data, such as instrument and test equipment calibration, software quality assurance, and hardware sampling.

It should be an ongoing process of the organization to identify appropriate environmental performance indicators that are objective, verifiable and reproducible. Indicators need to be relevant to the activities they are used to measure or monitor, as well as being practical, cost-effective and technologically viable.

Organizations, according to ISO 14001, must establish and maintain a procedure for periodically evaluating compliance with relevant environmental legislation and regulations as part of the monitoring program.

4.4.2 Auditing an Environmental Management System

Periodic audits of the system need to be conducted to determine conformance to requirements and agreements. The frequency of audits will be guided by the results of previous evaluations and the significance of potential environmental impacts.

Audits may be performed by internal personnel or selected external parties as long as the auditors are properly trained and can perform objectively and impartially.

Audit results need to be reported to those individuals within and outside the organization who are responsible for responding to adverse findings or are obligated by regulation or requirement to review the reports.

The environmental management specification, ISO 14001, defines an audit as a

> systematic and documented verification process to objectively obtain and evaluate evidence to determine whether an organization's Environmental Management System conforms to the...audit criteria set by the organization, and to communicate the results of this process to management.

The section of this primer titled Systems Checklist provides a listing of possible audit questions by area of evaluation. The questions are taken from ISO 14000 as issues to be considered when developing the key elements of the environmental management system. The same questions could result in effective and efficient audits of the system and performance.

There are a number of standards within ISO 14000 on auditing. Each standard provides guidance either on the performance of environmental audits, performance of environmental management system audits, the mechanisms for audits, or on the necessary qualifications of auditors. Here's the list of ISO 14000 audit standards and what they concern. Each begins with an introduction, followed by scope, normative references and definitions.

ISO 14010 *Guidelines for Environmental Auditing—General Principles*

This standard identifies the requirements for an environmental audit, gives the general principles of an environmental audit, and provides a description of the types of environmental audits.

ISO 14011/1 *Guidelines for Environmental Auditing—Audit Procedures*

This is a three part standard. Part 1 is on auditing an environmental management system. It discusses the objectives of the audit and associated roles and responsibilities. It also provides guidance on auditing and how to effectively completing an audit. Part 2 is about environmental compliance audits. And Part 3 is about auditing an environmental statement, known in the United States as an environmental impact statement, required by the U.S. Environmental Protection Agency for any major project.

ISO 14012 *Guidelines for Environmental Auditing—Criteria for Environmental Auditors*

This standard deals with the necessary education and work experience for an effective environmental auditor. It discusses the training an auditor should have, above and beyond education and experience.

Guidance is given on objective evidence which should be obtained about the auditor's education, experience, and training; the key personal attributes and skills of a successful environmental auditor; the qualifications required of a lead auditor; the maintenance of competence as an auditor; the meaning of "due professional care"; and guidance on language.

This standard has annexes on how to evaluate the qualifications of environmental auditors and on the environmental auditor registration body.

Other environmental auditing standards include:

ISO 14013 *Guidelines on Environmental Auditing—Management of Environmental Management System Audit Programs*

ISO 14014 *Guidelines for Initial Environmental Reviews*

ISO 14015 *Guidelines for Environmental Site Assessments*

In the United States, the American Society of Testing and Materials (ASTM) is working on auditing standards of their own: ASTM PS11, *Practice for Environmental Compliance Audits*, and ASTM PS12, *Guide for the Study and Evaluation of an Organization's Environmental Management System*. The National Science Foundation is working on an environmental auditing standard they propose to outrank ISO 14000, though that is very doubtful—NSF International 110, *Guiding Principles and General Requirements for Environmental Management Systems*.

British Standard 7750 describes audits as assessments of the effectiveness of the environmental management system as well as the achievement of environmental objectives. It recommends that all parts of the organization be audited at least every three years. Parts of the organization having a particular potential to cause environmental harm or damage should be audited at least once a year.

The primary function of an audit is to assess compliance and the effectiveness of previous corrective actions. Audits may also suggest remedial measures to overcome environmental problems. It may be of benefit for an organization to instill the habit of self-assessment by line management.

Audits are required by ISO 9000. Internal quality audits are conducted to verify whether activities and related results comply with planned arrangements and to check the effectiveness of the management system. They are to be scheduled on the basis of the status and importance of the activity to be audited. Results of internal audits are integral to management reviews. Combining an audit with an appraisal can take an organization beyond the traditional boundaries of auditing compliance and can appraise effectiveness. Audits determine implementation of requirements while appraisals determine performance.

There are certain qualities an auditor should have to be effective and efficient. First, an interest in researching situations to find out what was done, who did it, and why—being inquisitive. The talents of an auditor depend on experience in auditing and being audited as well. An auditor needs to be curious, cordial, comprehending, competent and courteous.

It's reported that there are 10 issues most commonly identified by ISO 9000 audits. Chances are, the same issues will surface during ISO 14000 management system audits. Here's a summary.

System Element	%	Common Issues
Document Control	20	·Documentation for a specific decision-making/implementation process lacking ·Documentation outdated and not available at relevant employee locations ·Lack of non-falsifiable way of assuring that procedures and manuals actually followed at all times
Design Control	12	·Little/no documentation for engineering calculations and design assumptions ·Outdated versions of engineering drawings not reliably eliminated from circulation ·No adequate proof as to when certain changes/modifications occurred
Purchasing	10	·Insufficient documentation regarding criteria used in purchasing process and lack of adherence to them ·Insufficient validation that criteria used actually correlates to technical and quality requirements and specifications ·Insufficient validation that purchasing personnel sufficiently trained to make certain decisions ·Insufficient data regarding ability (and track record) of vendors and subcontractors to meet contractual requirements
Inspection & Testing	10	·Insufficient documentation that no incoming product was released before inspected or verified to conformity ·Insufficient documentation as to inspection criteria to be utilized
Process Control	8	·Insufficient work instructions for employees and inadequate process monitoring during production and installation ·Lack of specific criteria for workmanship in written form or by representative samples · Special and unique/proprietary processes not documented/validated sufficiently
Inspection, and	8	·Insufficient records regarding timing and completeness of measuring control, calibration, and maintenance of measuring and test equipment ·Inadequately defined procedures for testing and calibration ·Insufficient assessment of causes in case equipment is found out of calibration and what remedial actions were taken ·Insufficient environmental controls to assure fitness for use ·Inadequate safeguarding of facilities, hardware and software from any action that would invalidate calibration
Contract Review	6	·Inability to demonstrate each contract was reviewed to assure capability to meet contract requirements
Corrective Action	5	·Inadequate documentation to provide investigation for root cause, and corrective actions taken to eliminate problem ·Lack of preventive action to deal with potential significant risks ·Lack of ongoing controls to assure corrective actions implemented and remain effective
Management	4	·Inadequate verification that programs have actually been responsibility implemented ·Employees not given adequate resources and training to carry out functions to necessary extent
Quality Records	3	·Lack of maintenance and retrievability of quality records ·Lack of non-falsifiable proof that certain records were compiled on day/time indicated and by qualified and authorized personnel

ISO 14000 provides the following objectives of an audit:

1. Determine conformity of the environmental management system elements with specified requirements and identify nonconformity.

2. Determine the effectiveness of the system in meeting objectives.

3. Provide opportunity for system improvements.

4. Meet regulatory requirements and commitments.

5. Meet ISO registration requirements.

Anyone in the organization may perform audits, if they are qualified and authorized. They should be free from bias and from influences that could impair their objectivity. This is usually accomplished by using personnel independent of the area audited.

A typical audit usually involves ten basic steps:

Step 1 Schedule the audit to occur at the best possible time for evaluating the intended areas to be audited.

Step 2 Define the purpose and scope of the audit.

Step 3 Make pre-audit notifications as required.

Step 4 Prepare for the audit with an audit plan and, preferably, an audit checklist; conduct as much preliminary review of pertinent documents as you can.

Step 5 Hold an audit entrance briefing to introduce the auditor(s) and the audit objectives and to identify contacts and necessary resources for the audit.

Step 6 Perform the audit; collect objective evidence through interviews, reviews, examinations and observations.

Step 7 Hold an audit exit briefing to identify the preliminary results and findings of the audit and get clarifications or corrections, if needed.

Step 8 Document the results in an audit report.

Step 9 Present the report to responsible management, with distribution as required by procedure or as requested by senior management.

Step 10 Conduct follow up on any corrective or preventive actions for identified nonconformities or weaknesses.

> Reporting audit results to management is the most important and demanding part of your job as an internal auditor. The value of your work comes not from the gathering of information, but from your assessment and presentation of it. Management's awareness and acceptance of your conclusions and its prompt action in response to reported problems are your measures of success (Maniak 1990).

Each audit report may cost between $10,000 and $100,000 U.S. dollars. That means some attention to the quality of the report is in order. Here are some tips that will add value to your audit reports (Ibid.):

Keep in mind the primary purposes of an audit report: inform, persuade and get results. Follow the pattern of logic of your readers when writing the audit report. What is your conclusion? What supports it? Why is it important? What should the reader do? "The value of your audit report lies in its ability to influence readers to change." Use terms your readers understand. Esoteric or excessively technical language will turn readers away. Choose the information of greatest interest and use to the reader. Guide the reader through the report (Ibid.).

These are the quality standards for internal audit reports (Ibid.):

- **direct**
  - conclusive opening sentences
  - informative headings
  - presentation of main points first
- **succinct**
  - select items of greatest significance
  - summarize supporting data
- **appropriate**
  - know readers' interests
  - select proper emphasis
  - present only relevant valid information
- **persuasive**
  - be convincing
  - quantify the impacts of identified conditions
- **constructive (causes, not symptoms)**
- **results-oriented**
  - measurable recommendations
  - practical solutions
  - action taken by management
- **inviting (professional format and content)**

- **timely**
  - immediate presentation to line management
  - interim reporting of serious issues

"Managers demand concise presentations, yet they do not want a document so terse that it leaves important questions unanswered.... A well-written conclusion gives management the information, perspective and balance they need for decision-making." Avoid judgmental language and unsupported generalizations. Use factual, concrete language. Beware connotations; they may become accusations. State positive ideas in positive language. Provide balance and perspective. Use active verbs. Use your own words. Avoid filler phrases, expanded or redundant modifiers, drawn-out verbs, the passive voice ("removes action from sentences"), repetition, extra articles, and prepositions. "Give every word a purpose, and your writing will be concise.... Writers have a natural sense of ownership and pride of authorship. Written communication is more personal, more of a self-expression than other methods of presentation or other kinds of work" (Ibid.).

The audit report should include an executive summary. It needs to state the audit objective(s), scope and any pertinent background information. Audit opinion or conclusion will be described in sufficient depth as well as the results of the audit. If necessary, appendix additional information, such as exhibits or attachments with details to support conclusions.

"Give an inviting look to the body of the report by making it easy to follow and comfortable to absorb." Here is a ten-point checklist for audit report quality (Ibid.):

1. Are the appropriate sections included?
2. Is there a summary?
3. Does the outline of the report give the reader a quick overview of the content?

4. Are enough headings used to guide the reader?

5. Are titles written in a consistent style?

6. Is the format inviting?

7. Are the most important comments presented first?

8. Are related comments combined?

9. Is repetition avoided?

10. Is detail presented in exhibits or appendices?

Below are six steps to an effective audit conclusion. They are important for internal environmental management audits (Ibid.):

1. Answer the audit objective(s).

2. Choose concrete supporting information.

3. Use an appropriate and consistent tone.

4. Balance the positive and negative.

5. Offer perspective.

6. Comment on management's response.

Avoid using buzzwords such as, "generally improved," "some control deficiencies," "compromise integrity" (Ibid.).

The audit report can be started prior to the audit. You can arrange the format, references you used to understand the area you're auditing, expected distribution, and so forth. This helps reduce the cost and expedites completion of the audit. It also gives you the opportunity to record conclusions and supporting observations as you identify them. This practice can shorten audit time by up to 25 %. It is important to issue the results of audits as soon as possible for effective response to problems identified.

> Audits should be performed to determine whether various elements within a quality management system are efficient and effective to satisfy customer needs and achieve quality objectives. For this purpose, appropriate quality requirements covering the system, product, service, and process should be documented and maintained (Stewart, Mauch and Straka 1994).

The best thing an auditor can do is to find that your problem areas have already been identified, especially the significant adverse conditions, and that you've fixed or are working on corrective and preventive actions. If the auditor is from your regulator and finds a significant condition, you will likely be reprimanded for waiting for the regulator to find it. In the eyes of the U.S. Nuclear Regulatory Commission as well as those of the U.S. Department of Energy, you should know your management systems well enough to spot adverse conditions before an outside audit does. Failure to do so is an indicator of production before safety and may be justification for shutting down your operations until you've learned how to manage.

4.5 Environmental Management Reviews and Improvements

Principle 5:

> **An organization should review and continually improve its environmental management system, with the objective of improving its overall environmental performance.**

The last section of ISO 14000 recommends a continual improvement process be applied to the environmental management system, starting with measuring and monitoring. There are three elements discussed:

- Review of the Environmental Management System
- Corrective and Preventive Action
- Continual Improvement

4.5.1 Reviewing an Environmental Management System

Management needs to review the system at appropriate intervals to ensure its continuing suitability and effectiveness. Its review should be broad in scope, yet in-depth enough to address all environmental dimensions of operations, the impact of operations on financial performance, and the impact of operations on the competitive position of the organization.

A number of sources of information need to be considered in the reviews:

- Environmental objectives, targets and performance
- Audit findings
- Evaluations of policy and effectiveness
- Changes in legislation
- Changes in expectations or requirements
- Organizational or operational changes
- New technologies
- Lessons learned
- Market preferences

British Standard 7750 describes reviews as checks on the continuing relevance of the environmental policy; updates on the evaluation of environmental impacts; and checks on the efficacy of audits, and follow up actions. The scope of the review should include the entire organization and its activities, products and services.

Reviews need to go beyond just compliance with policy. For example, if the review is of the environmental management system applied to product design, it should examine the extent to which the system was implemented, the effectiveness of the system in ensuring design according to objectives, and whether the objectives should be changed or modified.

According to BS 7750, these issues should be addressed in the review process as a minimum:

- recommendations in audit reports
- continuing suitability of the environmental policy
- emerging environmental concerns
- developing environmental issues
- potential regulatory developments
- concerns of interested parties
- market pressures
- changes in activities
- changes in the sensitivity of the environment

Reports resulting from environmental management system reviews need to make it clear why they were conducted. Any findings, conclusions or recommendations reached need to be documented. Senior management must respond to the identification of any significant adverse condition. It must take the necessary corrective or preventive action.

With management reviews, look beyond simple compliance. Examine your environmental management system like a physician would perform an annual health examination. Too much flab? Need more muscle? Need new specialized exercises? Any hidden performance problems? Any symptoms of adverse internal conditions? Problems with intake? Problems with output? Problems with wastes? New handicaps? Have conditions improved since the last checkup? Need for a change in treatment? Need to extend any prescription? Do existing records reflect the true current state of health?

4.5.2 Correcting and Preventing Environmental Impacts

It is important to document audit and review findings, conclusions and recommendations and to identify needed preventive or corrective actions. Any actions taken need follow up to determine effectiveness.

Requirements for documenting and correcting environmental impacts in ISO 14001 include the following:

- Establish and maintain procedures for handling and investigating nonconformance.
- Define responsibility and authority.
- Take action to mitigate impacts and to initiate corrective and preventive action.
- Take action commensurate with the environmental impact encountered.

Incidents of noncompliance with requirements may happen suddenly or accidentally, or they may be conditions that last for a period of time, according to BS 7750. They may result from deficiencies or failures in equipment, from human error, or from deficiencies in the environmental management system itself. Investigations of noncompliances need to determine and identify the causal mechanisms and the predisposing factors within the environmental management system.

Corrective action resulting from the investigation should involve restoring compliance as quickly as practicable. Recurrence needs to be prevented. Any adverse environmental impacts need to be mitigated. And the effectiveness of the actions needs to be fully demonstrated.

Human error is never the root cause of a noncompliance. If an error by an individual can result in a significant impact on the environment, then management needs to erect sufficient barriers so that one wrong action is mitigated from harming the environment.

Bob Moody, with LRL Energy Services, Inc., formerly with the Institute for Nuclear Power Operations (INPO), has studied human performance in the nuclear power industry for years and is convinced that human error can be a root cause. He cites the example of an employee who has been trained and reminded to wear a hard hat where required but who refuses to and, subsequently, suffers a head injury on the job. Bob believes the human error perpetrated by the individual is the root cause of the injury. I agree it may be the direct cause, and certainly is a contributing cause, but the root cause lies deeper. We have to answer why the behavior was allowed to occur.

Every root cause lies with some deficiency in the management systems or controls. People make mistakes for any number of reasons. Machines fail. Equipment fails. Processes leak. In any case, management is to blame for failing to plan ahead and be prepared to accommodate errors and failures without undue risk to the environment.

There are a number of techniques available for root cause determination and analysis. All appear to trace back to the Federal Aviation Agency's investigation and analysis tools used following aviation incidents and accidents. The Management Oversight and Risk Tree technique provides the most comprehensive root cause analysis system available, in the my opinion.

It always costs more to correct something than to prevent it ahead of time. More and more in the United States, prevention is taking precedence over correction. Even the U.S. Nuclear Regulatory Commission is paying more attention to preventive maintenance than they did originally. In most cases, preventive actions are fairly simple, fairly inexpensive (always a determining factor), and usually effective in avoiding catastrophes or significant impacts on the environment.

4.5.3 Continual Improvement of Environmental Management

Continual improvement as an ongoing process is embodied in the environmental management system. The environmental performance of the organization needs to be looked at and evaluated continuously to recognize and assure improvements are sought and implemented. ISO 14000 recommends that the continual improvement process should:

- Identify areas of opportunity for improvement,
- Determine the root cause of significant deficiencies,
- Correct the root cause and prevent its recurrence,
- Verify effectiveness of actions taken,
- Document changes resulting from process improvements, and
- Compare performance against objectives and targets.

The Systems Checklist in this primer lists questions that may be useful in conducting reviews, evaluating effectiveness of corrective and preventive actions, and seeking continual improvement. They are issues to be considered in developing the environmental management system, according to ISO 14000, but should also serve well after the system is up and running.

The time has now come to explore the details of IS0 14001.

ENDNOTES

1. Section 4 of ISO 14000 identifies all five principles. Each section related to a certain principle also includes the principle at the beginning of the section.

2. The word "paradigm," comes from the Greek. Originally a scientific term, it now means a model, a theory, perception, assumption, or a frame of reference. It is the way each of us sees the world in terms of perceiving, understanding and interpreting.

3. A friend of mine, Mark McClure, proposes you imagine a **sphere of concern** and a **sphere of influence**, so that you see them in the truest light of the four known dimensions.

4. Martin Buber in Bell, Chip R. 1994. *Customers as Partners: Building Relationships that Last a Lifetime.*

5. Fellers, Gary. *The Deming Vision.*

6. Environmental objectives are those overall goals an organization sets for itself to achieve, preferably with quantifiable measurements. Environmental targets are those measurable outputs relating to an organization's control of environmental impacts.

7. Quality due to product design involves those product design features that influence intended performance plus those that influence robust performance under variable conditions of production or use.

8. Pollution prevention involves use of processes, practices, materials, products, and energy that avoid or reduce the creation of pollution and waste. Prevention may necessitate recycling, process changes, control mechanisms, energy, and resource efficiency, or material substitution. There need to be objectives and targets for pollution prevention and waste reduction.

9. Interested parties are stakeholders, defined by ISO 9000: An individual or group of individuals with a common interest in the performance of the supplier organization and the environment in which it operates (customers, employees, owners, subsuppliers, and society).

10. Including those containing a subjective element.

11. Sibson, Robert E. 1994. *Maximizing Employee Productivity, a Manager's Guide*. Amacom Books; New York, NY.

12. I strongly advise you to assign each responsibility and associated authority to a specific individual instead of to a group of individuals. If the Board of Directors is responsible for an action, then the Chair of the Board should be held accountable. It has been learned from experience that shared responsibility means no responsibility, no accountability.

13. I also recommend the responsibility for the environmental policy be assigned to and accepted by the most senior management individual. If your organization has a Chief Environmental Manager or equivalent, this individual should also sign off to document and display commitment. But policies are high level documents and must be bought in to and produced by high level management.

14. See the two previous notes.

15. Each Manager should have specific areas of responsibility.

16. See the earlier footnotes on individual accountability. Also, each employee with responsibility for environmental performance should be responsible for ensuring continual improvement.

17. A new form of advertising is the "infomercial." This is usually a videotape for airing that promotes a product or service in a simulated situation that is intended as both a market for the

product/service and that shows conditions when the product/service works best.

18. The identification of records allows for the traceability required by ISO 14001.

19. Important to the storage of records is protection against damage, deterioration and loss.

CHAPTER 5

Into the Heart of ISO 14001

This chapter of the primer gives a detailed overview of the ISO 14001 environmental management system specification and guidelines. The salient and useful points from the standard are presented, and other advice has been added where pertinent.

0 Introduction

Increasingly stringent legislation, development of economic policies and measures to foster environmental protection, and the growing concern of interested parties about sustainable development—all are causing organizations to control the impact of their activities.

Reviews and audits to assess environmental performance may be less than sufficient. Customers and other stakeholders want assurance your organization will continue to meet requirements. To be effective, you need to work inside a structured management system, integrated into your overall management philosophy.

International environmental management standards intend to provide the elements of effective management systems to assist in achieving environmental and economic goals. They do so without creating non-tariff trade barriers or interfering with an organization's legal obligations.

This standard, ISO 14001, specifies environmental management system elements applicable to all types and sizes of organizations under diverse geographical, cultural and social conditions. Success depends on commitment from all levels of the organization. There must be demonstrated dedication to establishing and assessing effectiveness of environmental policy, objectives and procedures, and then to achieving conformance and demonstrating it to others. The aim of ISO 14001 is to support environmental protection in balance with socio-economic needs.

The important distinction between ISO 14001 and ISO 14000 is that ISO 14001 is the specification, describing the core elements for certification or self-declaration of an environmental management system, while ISO 14000 is a non-certifiable guidance standard. The specification is written in prescriptive language and contains only those system elements that may be objectively audited. This specification, however, establishes no absolute requirements for environmental performance beyond commitment.

Adopting ISO 14000 will only *promote* optimal environmental outcomes. To *achieve* environmental objectives, the environmental management system should encourage consideration of implementation of best state-of-the-art technology where available and economically viable.

The specification shares common management system principles with the ISO 9000 series of quality system standards. Organizations may elect to use an existing management system consistent with ISO 9000 as a basis for their environmental management systems. However, the application of various elements of the management system may differ due to different purposes and different interested parties.

Quality management systems deal with customer needs while environmental management systems address the needs of a broad range of interested parties and the evolving needs of society for environmental protection.

This section of the chapter summarizes the core elements of ISO 14001 and gives tips and suggestions on how best to comply.

The Table of Contents of ISO 14001 appears below:

1 Scope

The ISO 14001 standard specifies core requirements for environmental management systems. It does so without stating specific environmental performance criteria. It requires organizations to formulate policy and objectives taking into account legislative requirements and information about significant environmental impacts.

The ISO 14001 standard applies only to those environmental effects over which an organization has control and over which it can be expected to have an influence.

This specification applies to any organization with a desire to:

- implement, maintain and improve an environmental management system, **and/or**
- assure itself of conformance with policy, **and/or**
- demonstrate conformance to others, **and/or**
- seek certification/registration by an external organization, **and/or**
- make a self determination and declaration of conformance with ISO 14001.

All of the elements of ISO 14001 are to be incorporated into an environmental management system in a graded approach. The exact extent of incorporation depends on factors such as environmental policy, the nature of an organization's activities, and the conditions under which the organization operates.

Organizations have the freedom and flexibility to define their boundaries. They may choose to implement ISO 14001 throughout the entire organization. They may only apply it to specific operating units or activities.

An annex to ISO 14001 provides informative guidance on use of the specification. The annex is intended to clarify the core elements and avoid misinterpretation without adding to or subtracting from the content of the specification. According to the annex, ISO 14001 is based on the concept that the organization will periodically review and evaluate its environmental management system to identify opportunities for improvements in environmental performance.

The level of detail and complexity of the environmental management system should be driven by the size and nature of the organization. Other considerations include the extent of documentation and availability of resources, especially for small and medium-sized enterprises.

Remember, always ask yourself and your customers: Is it cost-effective? Does it add value? Do the benefits outweigh the risks?

2 References

The ISO 9000 series and ISO 14000

Refer to the list of standards in the ISO 14000 series earlier in this primer.

The ISO 9000 series contains these primary standards:

> In the United States, there is: ANSI/ASQC Q9000-1, *Quality Management and Quality Assurance Standards—Guidelines for Selection and Use*

ISO 9000-1 *Quality management and quality assurance—Part 1: Generic guidelines*

ISO 9000-2 *Quality management and quality assurance standards—Part 2: Generic guidelines for the application of ISO 9001, ISO 9002 and ISO 9003*

ISO 9000-3 *Quality management and quality assurance standards—Part 3: Guidelines for the application of ISO 9001 to the development, supply and maintenance of software*

ISO 9000-4 *Quality management and quality assurance standards—Part 4: Guide to dependability programme management*

ISO 9001 *Quality management and quality assurance standards—Design/development, production, installation and servicing*

ISO 9002 *Quality management and quality assurance standards—Production and installation*

ISO 9003 *Quality management and quality assurance standards—Final inspection and test*

U.S. equivalents:

ANSI/ASQC Q9001, *Quality Systems—Model for Quality Assurance in Design, Development, Production, Installation, and Servicing*

ANSI/ASQC Q9002, *Quality Systems—Model for Quality Assurance in Production, Installation, and Servicing*

ANSI/ASQC Q9003, *Quality Systems—Model for Quality Assurance in Final Inspection and Test*

ISO 9004-1 *Quality management and quality system elements—Part 1*

U.S. equivalent: ANSI/ASQC Q9004-1, *Quality Management and Quality Systems—Guidelines.*

ISO 9004-2 *Quality management and quality system elements—Part 2: Guidelines for services*

ISO 9004-3 *Quality management and quality system elements—Part 3: Guidelines for processed materials*

ISO 9004-4 *Quality management and quality system elements—Part 4: Guidelines for quality improvement*

ISO 9004-5 *Quality management and quality system elements—Part 5: Guidelines for project management*

ISO 9004-6 *Quality management and quality system elements—Part 6: Guidelines for quality plans*

ISO 9004-7 *Quality management and quality system elements—Part 7: Guidelines for configuration management*

ISO 10011-1 *Guidelines for auditing quality systems—Part 1: Auditing*

U.S. equivalent: ANSI/ASQC Q10011-1, *Guidelines for Auditing Quality Systems—Auditing.*

ISO 10011-2 *Guidelines for auditing quality systems—Part 2: Qualification criteria for quality system auditors*

U.S. equivalent: ANSI/ASQC Q10011-2, *Guidelines for Auditing Quality Systems—Qualification Criteria for Quality Systems Auditors.*

ISO 10011-3 *Guidelines for auditing quality systems—Part 3: Management of audit programmes*

U.S. equivalent: ANSI/ASQC Q10011-3, *Guidelines for Auditing Quality Systems—Management of Audit Programs.*

ISO 10012-1 *Quality assurance requirements for measuring equipment—Part 1: Metrological confirmation systems for measuring equipment*

ISO 10012-2 *Quality assurance requirements for measuring equipment—Part 2: Quality assurance*

ISO 10013 *Guidelines for developing quality manuals*

ISO 14014 *Guide to the economic effects of quality*

ISO 14015 *Continuing education and training guidelines*

ISO 8402-1 *Quality management and quality assurance—Vocabulary*

ISO Handbook—*Statistical methods*

3 Definitions

Definitions for ISO 14001 are the same as in ISO 14000, with the following exceptions:

Environmental management system audit—the ISO 14001 specification adds "and to communicate the results of the process to management."

Organization—*NOTE: for bodies or establishments with more than one operating unit, a single operating unit may be defined as an organization.*

Prevention of pollution—use of processes, practices, materials, products or energy that avoid or reduce the creation of pollution and waste. *NOTE: may include recycling, process changes, control mechanisms, energy and resource efficiency, and material substitution.*

4 Environmental Management System

4.0 General

Your organization is to establish and maintain an environmental management system with the core elements described in ISO 14001, as discussed in the Scope.

4.1 Environmental policy

Top management is to define the organization's environmental policy and ensure that it:

- is appropriate to the nature, scale and environmental impacts of its activities, products and services;
- includes a commitment to continual improvement and prevention of pollution;
- includes a commitment to comply with relevant environmental legislation and regulations and with other requirements under which the organization functions;
- provides the framework for setting and reviewing environmental objectives and targets;

Relate to Prep rev.

- is documented, implemented, maintained and communicated to all employees; and
- is available to the public.

The annex to ISO 14001 recommends that the environmental management system should enable to organization to:
- identify environmental aspects and impacts from past, exist ing, and planned activities, products and services.
- identify relevant legislative and regulatory requirements.
- identify priorities and set appropriate objectives and targets.

The annex to ISO 14001 recommends taking these additional steps for implementing an environmental management policy:
- establish a structure and program to implement the policy and achieve objectives and targets.
- facilitate planning, control, monitoring, corrective action, auditing, and review activities to ensure the policy is complied with and that the system remains appropriate.
- be capable of adapting to changing circumstances.

4.2 Planning

4.2.1 Environmental aspects

Your organization is to establish and maintain a procedure to identify your environmental aspects. The aspects of your activities, products and services that you can control and have an influence over are to be identified. The purpose is to determine those which have or can have significant impacts on the environment. Your organization must ensure these impacts are considered in setting your environmental objectives. You are expected to keep this information up-to-date.

> The annex to ISO 14001 recommends that the process take into account the cost and time of undertaking the analysis and the availability of reliable data. Information already developed for other purposes may be used.
>
> You should also take into account the degree of practical control you have over the environmental aspects being considered.
>
> The process should consider normal operating conditions, and shutdown and startup conditions, as well as realistic potential significant impacts associated with reasonably foreseeable or emergency situations. The control and influence over environmental aspects may vary significantly, depending on the market situation.

4.2.2 Legal and other requirements

Your organization is to establish and maintain a procedure to identify and have access to legal and other requirements to which you subscribe directly applicable to its environmental aspects.

> The annex to ISO 14001 identifies industry codes or practice, agreements with public authorities, and non-regulatory guidelines.

4.2.3 Objectives and targets

> The annex to ISO 14001 recommends that objectives be specific and targets be measurable, with preventive measures taken into account wherever possible. When considering technological options, you may want to consider the use of best available technology if cost-effective and appropriate.

Your organization will establish and maintain documented environmental objectives and targets at each relevant function and level within the organization. You are expected to consider relevant legal and other requirements, significant environmental aspects and technological options. You are also expected to consider financial, operational and business requirements as well as the views of interested parties. Objectives and targets must be consistent with your environmental policy, including your commitment to pollution prevention.

4.2.4 Environmental management programs

Your organization is to establish and maintain programs for achieving environmental objectives and targets, including: designation of responsibility for achieving objectives and targets at each relevant function and level of the organization **and** the means and time frame by which they are to be achieved

If appropriate, amend your programs to ensure that environmental management will also apply to new developments and new or modified activities, products, and services.

According to the annex to ISO 14001, the creation and use of this program is a key element to successful implementation of an environmental management system. The program should describe how targets will be achieved, with time scales, and subdivided to address specific elements of operations. The program should include an environmental review for new activities. It may also include consideration of planning, design, production, marketing, and disposal stages for both current and new products, services or activities. This may involve addressing design, materials, production processes, use, and ultimate disposal. For installations or significant modifications of processes, the program may also need to address construction, commissioning, operation, and decommissioning.

4.3 Implementation and operation

4.3.1 Structure and responsibility

Roles, responsibilities and authority are to be defined, documented and communicated to facilitate effective environmental management. Management is expected to provide resources essential to the implementation and control of the environmental management system. Resources include people, specialized skills, technologies and finances.

Top management will appoint a specific management representative or representatives with defined roles, responsibilities and authority irrespective of other responsibilities. The representative(s) is accountable for ensuring environmental management system requirements are established, implemented and maintained in accordance with ISO 14001. Management representatives will report on the performance of the system to top management for review and as a basis for improvement of the system. In small businesses, there is usually a single management representative who performs these services part-time while holding a separate responsible position.

> The annex of ISO 14001 suggests that environmental responsibilities be seen as including all areas of an organization and the commitment of all employees, from the highest level down through the designated management representative(s) to each worker. It is top management's responsibility to provide appropriate resources to empower employees.
>
> Empowerment, in my opinion, involves sufficient authority to act on responsibilities, adequate training and competence, the right tools and procedures at the right time, and freedom from harassment or fear of failure.

4.3.2 Training, awareness and competence

Your organization has to identify training needs for everyone who may create a significant impact on the environment. You have to require and provide appropriate training. You are expected to establish and maintain procedures to make employees or members at all levels aware of:

- the importance of conformance with environmental policy, procedures and the requirements of the environmental management system;
- the significant environmental impacts, actual or potential, of their work activities and the environmental benefits of improved personal performance;
- their roles and responsibilities in achieving conformance with policy, procedures and requirements, including emergency preparedness and response requirements; and
- the potential consequences of departure from specific operating procedures.

Personnel performing tasks which can cause significant environmental impacts must be competent on the basis of education, appropriate training and experience.

According to the annex to ISO 14001, management is to determine the level of experience, competence and training necessary to ensure the capabilities of personnel carrying out specialized environmental management functions.

4.3.3 Communication

Your organization is to establish and maintain procedures for internal communications. Procedures need to deal with communication between various functions and levels of the organization. You also need procedures for receiving, documenting and responding to relevant communication from external interested parties. External parties need to be kept informed regarding environmental aspects and your environmental management system. You are to consider processes for external communication on significant environmental aspects and record your decisions.

The communication procedures, according to the ISO 14001 annex, may include a dialogue with interested parties on their concerns. In some circumstances, responses to concerns may involve information about the environmental impacts of an organization's operations. The procedures should also address necessary communications with public authorities on emergency planning.

4.3.4 Environmental management system documentation

Your organization is to establish and maintain information in paper or electronic form to describe the core elements of your management system. You are also to describe their interaction and provide direction to related documentation.

The annex of ISO 14001 recommends that documentation provide direction on where to obtain more detailed information. This usually is for the operation of specific parts of the system. Documentation may be integrated and shared with that of other systems. Related documentation could include:

- process information, **and/or**
- organizational charts, **and/or**
- internal standards and operational procedures, **and/or**
- site emergency plans.

4.3.5 Document control

Your organization is to establish and maintain procedures for controlling all documents required by this standard to ensure:

- they can be located;
- they are periodically reviewed, revised as necessary, and approved for adequacy by authorized personnel;
- the current versions of relevant documents are available at all locations where operations essential to the effective functioning of the system are performed;
- obsolete documents are promptly removed from all points of issue and points of use or otherwise assured against unintended use; **and**
- any obsolete documents retained for legal and/or archival purposes are suitably identified.

Documentation needs to be legible, dated (with dates of revision) and readily identifiable. It needs to be maintained in an orderly manner and retained for a specified period. Procedures and responsibilities must be established and maintained concerning the creation and modification of the various types of documents.

The intent of ISO 14001, according to its annex, is to ensure that organizations create and maintain documents in a manner sufficient to implement the environmental management system. The primary focus, however, should be on effective implementation of the system and environmental performance instead of on a complex document control system.

4.3.6 Operational control

Your organization is expected to identify those operations and activities that are associated with identified significant environmental impacts. These are the impacts which fall within the scope of your policy, objectives and targets. You are expected to plan your activities, including maintenance, in order to ensure that they are carried out under specified conditions:

- establish and maintain documented procedures to cover situations where their absence could lead to deviations from policy, objectives or targets;
- stipulate operating criteria in the procedures;
- establish and maintain procedures related to significant environmental aspects of goods and services used; and
- communicate on relevant procedures and requirements to suppliers and contractors.

4.3.7 Emergency preparedness and response

Your organization is expected to establish and maintain procedures to identify the potential for and to respond to accidents and emergency situations. These procedures are to prevent and mitigate environmental impacts that may be associated with the emergencies. Review and revise emergency preparedness and response procedures when necessary to keep them current. Especially review procedures after an accident or emergency situation. You are also expected to test the procedures periodically.

4.4 Checking and corrective action

4.4.1 Monitoring and measurement

Your organization needs to establish and maintain procedures to monitor and measure key characteristics of operations and activities. Both evaluations are to be accomplished on a regular basis. Characteristics in question are those that can have a significant impact on the environment.

You are to record information to track performance, relevant operational controls and conformance with the objectives and targets. Monitoring equipment is to be calibrated and maintained and records retained. You are to establish and maintain a procedure for periodically evaluating compliance with relevant environmental legislation and regulations.

4.4.2 Nonconformance and corrective and preventive action

Your organization is to establish and maintain procedures and define responsibilities and authority for handling and investigating nonconforming conditions. Procedures need to address taking action to mitigate impacts caused by nonconformances.

Procedures need to be in place for initiating corrective and preventive action. Any corrective or preventive action taken to eliminate the causes of actual and potential nonconformances must be appropriate to the magnitude of problems and commensurate with the environmental impact encountered. Remember to implement and record any changes in the documented procedures resulting from corrective and preventive action.

The annex of ISO 14001 recommends four basic elements be included in the procedures for investigating nonconformances and corrective actions:

- identify the cause of the nonconformance,
- identify and implement necessary corrective action,
- implement or modify controls to avoid repetition, and
- record any changes in procedures resulting from corrective action.

Those actions may be accomplished rapidly and with a minimum level of formal planning or may be more complex and of a longer duration.

Implementing or modifying controls to avoid recurrence of a significant nonconformance means knowing the root cause of the problem through root cause analysis using an accepted method of evaluation.

The direct cause of a nonconformance is usually different than the root cause; same for contributing causes; e.g., what appears to be a personnel error in failure to follow procedures may be connected with *why* the person was allowed to deviate from procedures.

Ask questions. Was it lack of training? Obsolete copies of the procedure at the work location? Productivity push by supervision? Lack of needed resources? Problems with the procedure flow? Direction given by supervision to do the task a different way? Employee health problems?

Ask "why" and "how come" until there is no "why" or "how come" left to ask. Direct derivation is a good starting point, but structured analysis is usually more beneficial to prevent recurrence of significant nonconformances.

4.4.3 Records

Your organization is to establish and maintain procedures for the identification, maintenance, and disposition of environmental records. At a minimum records will include training records and the results of audits and reviews. They are to demonstrate conformance to the requirements of ISO 14001.

Environmental records must be legible, identifiable and traceable to the activity, product or service involved.

Records are to be stored and maintained in such a way that they are readily retrievable and protected against damage, deterioration or loss. Their retention times are to be established and recorded.

According to the ISO 14001 annex, these records management procedures should focus on records needed for implementation and operation of the environmental management system and should record the extent to which planned objectives and targets are met. Environmental management records may include:

- incident reports
- complaints and responses
- contractor and supplier information
- inspection, maintenance and calibration data
- process information
- product information
- training records
- audit results
- records of significant environmental impacts
- information on applicable environmental laws and other requirements
- emergency preparedness and response
- management reviews
- pertinent confidential business information

4.4.4 Environmental management system audit

Your audit program and schedule will be based on the environmental importance of the activity concerned and the results of previous audits. In order to be comprehensive, audit procedures need to cover the audit scope, frequency and methodologies, as well as the responsibilities and requirements for conducting audits and reporting results.

An audit program should cover at least six items, according to the annex of ISO 14001:

- activities and areas to be audited
- frequency of audits
- responsibilities associated with managing and conducting audits
- communication of audit findings
- auditor competence
- how audits are conducted

Audits may be conducted by personnel within the organization or external to the organization as long as they are competent, impartial, and objective.

Audits of companies, vendors or government organizations generally require that the auditor have no direct responsibility for the work or area audited, to ensure impartiality. Most auditors are trained in auditing practices, interview techniques, investigation tools, and sometimes root cause analysis.

For ISO 9000 auditors, training, and even certification, is offered by a number of organizations, and it is expected the same will become true for ISO 14000 auditors. The internal auditor should be trained in ISO 14000 auditing techniques and practices to become the organization's lead auditor. Then anyone in the organization with the necessary competence can participate under the lead auditor. Team audits, although more expensive than a sole auditor, usually yield more useful results since there are several individual opinions, paradigms and backgrounds involved.

4.5 Management review

Top management is expected to review the environmental management system at determined intervals. The purpose of management review is to ensure the system's continuing stability, adequacy and effectiveness. The management review process will ensure that necessary information is collected to allow management to carry out the evaluation. Reviews are to be documented. They are to address the need for changes to policy, objectives and elements of the system in light of audit results, changing circumstances and the commitment to continual improvement.

The ISO 14001 annex states that in order to maintain continual improvement, the organization's management should review and evaluate the environmental management system at defined intervals. The same applies to maintaining the suitability and effectiveness of the system. This review should be comprehensive even when it takes place over time and only addresses certain components at a time. The review needs to be carried out by the level of management that defined them and should include the:

- audit results,
- extent to which objectives and targets are met,
- suitability of the environmental management system in relation to changing conditions and information, and
- concerns of interested parties.

The documented results of reviews should identify observations, conclusions and recommendations.

Chapter 6

Making ISO 14000 a Reality

Focus on Process

Here's a way to remember what ISO (and Total Quality) management systems are all about:

Plan for failure
Educate for success
Organize the processes
Practice your policy
Lead to improve
Evaluate your efforts

Delegate the authority
Empower your people
Trust your instincts
Encourage other opinions
Respect your customers
Motivate your suppliers
Integrate your resources
Negotiate with flair
Engage your fears

Quit managing failure
Utilize your enemies
Activate your imagination
Love your work
Imitate the masters
Try something new
Yell for help

Of Kourse!

There is no one way to systematically and methodically manage environmental aspects, but the ISO 14000 path will lead you in the right direction. It is no more than common sense for best management practices.

ISO 14000 is a concerted international attempt to improve the quality of life with minimum effect on the environment, to transform with the minimum of alteration or distortion.

The future ISO 14000 registration process is likely to be similar to that of ISO 9000. No particular process will be required by ISO 14000, but there will be guidance on becoming registered in the standard. There are, however, a number of general steps an organization can take. Other more detailed steps will depend on the maturity of the organization's environmental management system, type of industry and specific requirements of the registrar.

General and typical steps to become ISO 14000 registered are described below. The same steps are helpful in tailoring your environmental management system to ISO 14000 without seeking registration.

STEP 1 Understand the standard, its impact on your organization, its requirements; determine the benefits of becoming registered.

STEP 2 Understand the registration process and decide whether to commit to becoming registered.

STEP 3 Authorize your Management Representative for the environmental management system to go forward with the process, with team members from each affected area of your organization, if possible; provide any needed training on the process.

STEP 4 Using input from your Management Representative, select an appropriate registering organization; check their waiting list; understand their process, expectations and requirements.

STEP 5 Perform internal audits of your environmental management system against each element of the standard; identify strengths and weaknesses; repeat audits where deficiencies are identified.

STEP 6 Correct deficiencies; conclude full compliance with the standard or develop a compliance plan to address discrepancies beyond the current capability to correct.

STEP 7 Prepare, or verify that you have clearly defined environmental management policies, objectives, targets, and procedures (usually in an Environmental Management System Manual or similar document); validate proper flowdown of regulatory and other commitments.

STEP 8 Again, correct deficiencies or document and implement a corrective action plan to resolve those that are beyond your current capability to address.

STEP 9 Apply for registration; meet with the registrar for indoctrination on your organization, its activities, products, and services; schedule the final audit (the registrar may request a pre-audit).

STEP 10 Expedite the final audit (say what you do and do what you say); accept the outcome (recommendation for or against registration); schedule correction of any identified deficiencies within 60 days and submit evidence of completed actions; expedite re-audit if major discrepancies are identified.

Your ISO 14000 certificate should be posted in a very conspicuous place for both your customers and your organization. If you chose only to follow the ISO 14000 registration process without certification, it would be of value to have a similar certificate or other type of award trophy created and displayed. If you receive high praise from a stakeholder or a customer, publicize it.

Developing, implementing and maintaining a sound environmental management system should receive serious publicity. It shows your commitment to continuous improvement in your impact and interface with the environment and ecosystem. It shows your concern for the protection of the health and safety of the public. It shows your voluntary obligation to the preservation of the environment for future generations.

If you decide to become ISO 14000 certified, or if you would like assistance from an ISO 9000 registrar who will become an ISO 14000 registrar, you should be able to find them through the contacts identified in the discussion on Scope earlier in this primer (ASQC, ANSI, ASTM, etc.; in addition, check the Resources in the Bibliography). Questions to consider when choosing a registrar include:

? Does the registrar have accreditation acceptable to your customer? Your regulator?

? Is the registrar's certificate recognized in each of the different countries where you expect to do business?

? Does the registrar have sufficient experience in your industry? Strong references?

? Does the registrar have agreements with other registrars in case you want to expand coverage?

? Is there a long waiting list?

? Are fees competitive?

? Are there any conditions on the registrar for use of its symbol/logo?

? What registry list(s) is published?

? What is your recourse if the registrar goes out of business? Loses accreditation?

? What about post-registration confirmation audits? Re-registration?

Making Environmental Management a Reality

Terrence Burton and John Moran, with the Galileo Electro-Optics Corporation, used fractal thinking to develop a model that could "function like a kaleidoscope" for reengineering their organization (Burton and Moran 1995). Galileo had to shift from supplying fiber- and electro-optic products to the U.S. Department of Defense, to new commercial markets. The company "recognized that the velocity and magnitude of change required to support a successful commercial transition exceeded our continuous improvement efforts and that focusing on the 'as is' was just not enough to make a difference in the global marketplace. Using fractal thinking, they accepted chaos as the root of their technology and moved out of their 'comfort zones' (Ibid.).

> "In an effort to maximize the performance of any organization, one might reengineer the traditional organization by transforming its structure from one that is department or problem oriented into one made up of holistic self-managed teams.... The team approach is a way of tapping into the dormant resources that are available, encouraging every employee to use their mind and take on ownership of their work." Chaos exists in business and in life. It is created by simple structured events. "Chaos is nothing more than unpredictable order with some limits" (Patterson and Mancini 1995).

Burton and Moran suggest a simple approach to creating your organization's vision statement:

We will be a better organization than we are today by....

Top management will personally lead our future-focused organization by....

Our employees, who are our most important asset, will....

We will continually strive to....

Our future-focused organization will have a social focus that will be an inspiration to the rest of our....

We will be the industry benchmark leader in the 21st century.

The Realities of Integrating Economics and Environment

> "We're a society of wasteful consumers. But employers and manufacturers have to foot the price of cleanups and pay tremendous fees to force government regulations.... Companies are naturally looking at their bottom line. It's about time people got together and found ways of protecting the bottom line but not at the expense of protecting the environment" (Marcinkowski, *San Diego Business Journal*, 1994).

> "There's a preponderance of toxic industries in poor communities of color...where there are some highly polluting industries. What are some of the strategies that we can use to achieve environmental justice and maintain a good economy?" (Takvorian, *San Diego Business Journal*, 1994).

President Bill Clinton has established the Council on Sustainable Development to explore and make policy recommendations on ways to integrate economic goals and environmental protection (Illman, *Chemical and Engineering News* 1994). The Administration is committed to an economic policy that incorporates environmental concerns and implements an environmental policy which strives to meet economic goals. The Council has 25 leaders from business, environmental firms and the government. They hope to recommend ways for the United States to fulfill commitments made at the Rio de Janeiro 1992 Summit, with a final report due the President by November 1995 (Miller, *Industry Week* 1994).

One of the Council's key recommendations is "the use of market incentives to achieve environmental goals wherever feasible," according to Chair Jonathan Lash, the President of the World Resources Institute. This recommendation "will lead to a significant change in the underlying approach to pollution control and environmental protection." The Council's recommendations place emphasis on performance-based regulations with companies allowed to devise their own compliance strategies (Ibid.).

Even Mikhail Gorbachev recognizes a need for balance between economics and environmental concerns. He heads the International Green Cross/Green Crescent organization, which is dedicated to stopping environmental damage and has spoken strongly against progress and technology that creates environmental hazards (Linden 1993).

Sustainable development can provide an economic advantage according to John Spisak, CEO of Industrial Compliance, a subsidiary of Southern Pacific Transportation: "CEOs must start to admit that the world is changing and that they are operating in a new arena where environmental problems will not go away. They have to begin to take the environment more seriously and begin to view environmental issues as a way to save dollars and preserve resources. It is the responsibility and obligation of industry to try and turn this around. CEOs need a willingness to take a new approach for the late '90s and beyond" (*Industry Week* 1994).

Look at how change to your production processes will improve the environment and lower your costs, compared to your competition's. "The net, from a true business standpoint, would almost always be an economic gain and an environmental advantage" (Ibid).

Know that every action has consequences, positive and negative. But without action, the world is doomed, this time with very personal consequences for you, your loved ones, your customers. Have the dream to go forward and do better. Many will follow.

ENDNOTES

1. Fractal geometry is accredited to Benoit B. Mandelbrot, who christened this new field of study in 1975. Fractal geometry has to do with mathematics, multiple iterations, and self-symmetry. The resulting infinitely complex image, if magnified, will eventually show the original image.

CHAPTER 7

The Future of ISO 14000

First, it's important to be aware of the current schedule for publishing the ISO 14000:

Environmental Management Systems

ISO 14000	Summer 1996
ISO 14001	Summer 1996

Environmental Auditing

ISO 14010	Summer 1996
ISO 14011/1	Summer 1996
ISO 14012	Summer 1996

Environmental Labeling

ISO 14020	Fall 1996
ISO 14021	Summer 1996

Life Cycle Analysis

ISO 14040	Summer 1996
ISO 14041	Spring 1997
ISO 14043	Spring 1997

The *Environmental Aspects in Product Standards Guide* is scheduled for publication in the Summer of 1996, slightly ahead of the environmental management system standards.

The Ciba-Geigy plant in Newport, Delaware, is using British Standard 7750 as a guide to prepare for ISO 14000 (Fairley and Roberts 1995). The status of Austria's Fachverband der Chemischen Industrie Österreichs (FCIO; Vienna) is the most rigorous and structured program (Ibid). The FCIO requires third-party independent verification of a company before they can use the FCIO "Care" logo. Care was pioneered in Europe by the UK's Chemical Industries Association (CIA; London) back in 1989. Care was started by FCIO in 1993 and to date, fewer than 20 of the 50 member companies that have applied have been granted the logo.

The U.S. Chemical Manufacturers Association has a similar program called "Responsible Care." Both care programs deal with due diligence and common sense environmental management and improvement.

The CIA is lobbying ISO and the British Standards Institute to accept its guidance document (Ibid.). The CIA wants its document used as the basis for developing an integrated, international standard on the environment, health and safety. It contends that the standard "should be incorporated into the structures already existing in BS 7750 and ISO 14001." This effort may affect the final release of ISO 14000 and delay release of its subsequent standards.

A separate organization, the Commission of the European Union's Eco-Management and Audit Scheme formally started up in April 1995 (Ehrle 1995). It requires that companies meet established environmental improvement targets, then have their environmental management systems verified by accredited, independent audit before they can use the "Emas" logo. The European Union is deliberating whether ISO 14000, BS 7750, or both will satisfy the Emas requirements. The main difference between them is public disclosure of environmental management information, especially audit reports,). The standards themselves have no requirement for disclosure, but Emas does so that public pressure can induce higher levels of environmental performance and enforcement. The European Union has formed a committee to write a "bridging document" to connect ISO 14000 and Emas requirements.

There is no doubt that ISO 14000 compliance will find its way into the United States fairly quickly. According to Gary Johnson of the EPA, the U.S. Environmental Protection Agency will probably reference the standards as voluntary indicators of systematic environmental management. There has been at least one Request for Proposal from the U.S. Department of Energy calling for compliance with ISO 14000.

The ISO 9000 registrars are looking forward to publication of ISO 14000. At least two I know of expect to offer environmental management system registration and third-party audits in the United States as soon as ISO 14000 is contractually required. Smithers Quality Assessments, Inc. is rapidly expanding in ISO 9000 registration as more U.S. organizations respond to contract requirements from overseas. The next step is ISO 14000 certification, with the need for a qualified registration body. Future ISO 14000 registration is a priority being eyed carefully by Underwriters Laboratories.

Appendix A

Environmental Management System Checklist

This portion of the primer lists the fundamental questions to ask and answer when developing an environmental management system. These questions are also useful in auditing the system and are a good guide to use for management reviews. Begin with your environmental management policy. Examine your management plans. Look closely at the elements of your management system. Check the "how to's" and analyze the "why's" of system design and implementation.

Each of the questions may be of value in establishing internal performance and evaluation criteria. It you expect to measure performance against set objectives, then make sure the objectives have been clearly identified.

Documentation is important as objective evidence for traceability, audit and review. But more importantly the documentation should reflect the positive as well as negative aspects of your environmental management system.

Anticipate answers before you develop a question or ask a question. Make evaluation and investigation scientific. Otherwise, data could produce inconclusive or inaccurate results.

Policy

1. Is there a documented environmental policy?[1]
2. Does the policy reflect appropriate values and guiding principles?
3. Has the policy been approved by the highest level of management (the Board of Directors)? Other governing body?
4. Has someone been delegated authority in writing to oversee policy implementation? To implement the policy?
5. Does the policy drive setting environmental objectives and targets?
6. Does the policy guide employees towards monitoring best available technology? Best management practices?
7. Does the policy commit to continual improvement?
8. Does the policy commit to meeting or exceeding legal requirements? Expectations of interested parties?

According to ISO 9000, there are a few other questions to be answered.

9. Do organization charts reflect the current organization?
10. Are responsibilities and authority of individuals who manage, perform, and verify the quality of work understood?
11. Is there a designated Management Representative? Are all others in the organization aware of the Representative's responsibilities, authority and the best way to communicate with the Representative? (The standard recommends that the name of the Management Representative be widely publicized and distributed.)

12. Are senior management reviews conducted as scheduled? Are reviews documented?

Environmental Aspects and Impacts

1. Have environmental aspects been identified for activities? Products? Services?

2. Do activities create a positive or negative change to the environment? Do products? Do services?

3. Does facility location organization require special environmental consideration?

4. Will intended changes or additions to activities alter environmental aspects and impacts? Will changes/additions to products? Will changes/additions to services?

5. Are there significant or severe potential environmental impacts should a process failure occur?

6. How frequent are there situations that could lead to significant/severe environmental impact?

7. What are the significant environmental impacts? (consider environmental aspects, likelihood, severity, and frequency)

When you analyze environmental aspects and potential impacts, ISO 9000 recommends you focus on design, assignment of activities and organizational interfaces.

8. Are there plans in place for each design and development activity? Do they identify responsibilities and authority?

9. Are product design inputs defined? Reviewed? Design differences resolved by the appropriate function(s)?

10. Are design outputs documented? Do they meet inputs? Are acceptance criteria specified?

11. Does the design conform to regulatory and other specified requirements?

12. Are the vital characteristics of safe function identified?

13. Are designs verified by competent personnel? Verified against inputs?

14. Are there design control measures in place? Are changes identified, documented, reviewed, and approved consistent with the original design?

15. Do responsible design and verification personnel have adequate resources?

16. Has the proper flow of technical information within the organization been defined and documented? Is the flow understood?

After examining system design ISO 9000 suggests you question your production process planning:

17. Does planning for production processes ensure operation under controlled conditions?

18. Are abnormal operating conditions and contingencies considered in the production planning?

19. What controls are applied to production processes? What approvals? What workmanship criteria? What process monitoring?

20. Have special processes been qualified? Are special process operators and equipment qualified? Records maintained?

21. Are special processes monitored on a continuous basis? Records maintained?

Environmental Objectives and Targets

1. Are the established environmental objectives and targets
   - within the context of the policy?
   - identified in terms of specific, measurable indicators?
   - identified by people responsible for achieving them?
2. Do they
   - reflect environmental aspects and significant environmental impacts?
   - consider views of interested parties?
3. Are objectives and targets regularly reviewed? Revised to reflect desired improvements in environmental performance?

Environmental Plans and Program

1. Is there an environmental management planning process?
2. Does the process involve all responsible parties?
3. Is the environmental management plan linked to policy? Objectives? Targets?
4. Is there a process for periodic plan reviews?
5. Is there a process for developing environmental management programs?
6. Do programs address resources, responsibility, timing, and priority?

7. Are programs fully integrated with policy? With the environmental management plan?

8. Are programs monitored on an ongoing basis as part of the operational review process?

Legal requirements and internal performance criteria should be questioned, as well:

9. Is a list maintained of laws and regulations that pertain to activities, products and services? Are authorizations, licenses and permits identified and tracked?

10. Have internal priorities and criteria been developed? Are they associated with external standards? Do they define required performance to fulfill policies, objectives and targets?

When it comes to planning, ISO 9000 adds questions of value to the environmental management system checklist. But it mentions one concerning the production process:

9. Is the production process planned? Under controlled conditions? Under emergency conditions?

Handling, storage, packaging, and delivery of product is discussed in ISO 9000. It would be wise to include these questions in your system checklist:

10. Are there means and methods in place to prevent damage to or deterioration of products?

11. Are there plans in place for periodic inspection of products when handled? Stored? Packaged? Shipped/delivered?

Human, Physical and Financial Resources

1. How are human, technical and financial resources identified and allocated to meet the systems' objectives and targets?

2. Is there a process to evaluate the environmental resource requirements (human, technical, financial) associated with capital projects?

3. Are there procedures to track costs and benefits (Return On Investment) of environmental activities?

One of the resources you need, according to ISO 9000, is a calibration program for measuring and test equipment. These questions should also be answered:

4. Has equipment for inspecting, measuring and testing product quality been identified? Is it calibrated and maintained?

5. Is the calibration traceable to national or other recognized standards?

6. Are there trained personnel responsible for calibrating equipment?

7. Are there documented calibration procedures?

8. Is the calibration status for each item uniquely identified?

9. Are calibration records kept up-to-date?

Training is a resource but one often overlooked in empowering people to totally manage the quality of their actions. The process for training should be questioned:

10. Have training needs been identified for personnel affecting environmental aspects?

11. Are people qualified based on education, training and experience?

12. Are training records on file?

13. Are qualifications evaluated periodically? Kept current? Adapted to changes in requirements, techniques and tools?

Organizational Alignment and Integration

1. Is the environmental management system integrated into the business management process?

2. Is there a process to resolve conflicts between environmental and other business objectives and priorities?

Accountability and Responsibility

1. Are key personnel responsibilities and accountability defined and documented for those who manage, perform and verify work affecting the environment?

2. Do these personnel

   - have sufficient training and resources for implementation?

   - initiate action to ensure compliance with policy?

   - anticipate, identify and record environmental problems?

   - initiate, recommend or provide solutions to problems?

   - verify implementation of solutions?

- control further activities until environmental deficiencies or unsatisfactory conditions are corrected?
- know how to act in emergencies?
- understand the consequences of noncompliance?
- understand their accountability?
- encourage voluntary action and initiatives?
- receive recognition and rewards for their performance?

Purchasing involves assignment of responsibilities to suppliers and subcontractors. In light of this, ISO 9000 recommends the following questions be answered:

3. What is the process for selecting suppliers? Is a list of acceptable suppliers maintained? What is the acceptance criteria?
4. Is information on purchase orders pertinent and complete? Is it current? Correct?
5. Are purchased products verified? On receipt? By source inspection?

Environmental Values

1. Does management establish, reinforce and communicate environmental values?
2. Do employees understand, accept and share the values?
3. Do the values motivate environmentally responsible actions?
4. Do performance reviews and rewards and recognition include environmental values?

Knowledge, Skills and Training

1. Are environmental training needs identified?
2. Are training needs of specific job functions analyzed?
3. Is the training program in place? Reviewed regularly?
4. Does the training process include documentation and evaluation?

Communication and Reporting

1. Is there a system for receiving and responding to concerns?
2. Is there a process for communicating on environmental policy and performance?
3. Are results from environmental management system audits and reviews communicated to all appropriate people?
4. Do internal communications support continual improvement on environmental issues?

Documentation

1. Are environmental management procedures identified? Documented? Communicated?
2. Is there sufficient evidence of the existence, implementation and maintenance of the environmental management system?
3. Is environmental management system documentation integrated with existing documentation?

Documentation and document control are important management issues, according to ISO 9000. These questions should also be answered:

4. Are procedures properly reviewed and approved? Are purchasing documents? Quality plans? Process controls? Audits?

5. Are current and correct documents available at the work locations? Do obsolete documents exist? Are they used?

6. Are changes to documents reviewed and approved consistent with the original issues?

7. Do personnel responsible for reviews have pertinent background information available?

8. Is there a controlled listing of documents and changes? Is there a procedure to assure obsolete documents are purged?

Traceability of products whose performance is important to the health and safety of the public or the protection of the environment is vital. If a catastrophe occurs because of product failure in service, then the history of the product, its production, inspection, design, and the procurement of its components is required if a root cause analysis is necessary to prevent recurrence. Thus, the following questions should be added to the system checklist:

9. Are the product and its vital components uniquely identified from concept to design? Through procurement and fabrication? Through production to installation or use?

Records and Information Management

1. Does the organization have access to environmental information it needs to manage effectively?

2. Is the organization capable of identifying and tracking key performance indicators for achieving environmental objectives?

3. Is there a records management system that makes information available to people when they need it?

Quality records are required by ISO 9000. It recommends answering the following questions:

4. Are records of product quality and activities affecting product quality and services maintained? Retained for specified intervals? Disposed of according to procedure?

5. Are records traceable to the activity, product or service they pertain to?

Emergency Preparedness and Response

1. Are there plans and procedures in place for potential emergencies and environmental incidents? Under normal and abnormal operating conditions?

2. Do emergency plans and procedures define roles, responsibilities and authority?

3. Do emergency plans and procedures provide details on emergency services? On actions taken in different types of emergencies? Have procedures been tested?

4. Do emergency procedures include information on hazardous materials likely to be encountered? Each material's potential impact on the environment? Measures to be taken in the event of accidental release?

5. Have emergency response personnel been properly trained? Tested for effectiveness?

6. Have alternative backup personnel been designated? Is the emergency response organization fully staffed at all times?

Measuring and Monitoring

1. Is environmental performance regularly monitored?

2. Have specific performance indicators been established that relate to the organization's objectives and targets?

3. Are the control processes in place adequate enough to regularly calibrate and sample measuring and monitoring equipment and systems?

Additional measuring and monitoring requirements are given in ISO 9000 for procurements. The following questions may be useful in the system checklist:

4. How are suppliers/subcontractors selected?

5. Is there a list of approved suppliers/subcontractors? Is it used?

6. Are purchased products/services verified for conformance? Are customer-supplied products/components?

7. Are special processes monitored continuously? (welding, brazing, heat treating, nondestructive examination—processes whose critical characteristics need to be observed or checked while in progress since inspection after the fact may be impractical or add no value)

8. Are internal audits scheduled and performed?

9. Are audit results communicated to personnel responsible for the areas evaluated? Responsible for corrective/preventive action?

10. Are audit findings responded to? Is corrective/preventive action taken?

Statistical quality control is a virtue, according to ISO 9000. Use of statistical controls may be advantageous in your environmental management system:

11. Are adequate statistical techniques available and used to verify process capability? Product criteria?

12. Are statistical techniques for product quality monitored and evaluated? Verified or validated?

Review

1. Is there periodic review of the Environmental Management System?
2. Are people involved in review of the Environmental Management System responsible for actions and follow up?
3. Are reviews communicated to interested parties? Other stakeholders?

Corrective and Preventive Action and Continual Improvement

1. Is there a process to identify areas for corrective and preventive action? For continual improvement?
2. Are the effectiveness and timeliness of corrective and preventive actions verified? Are trends documented?

Nonconformance control is key to corrective/preventive action as well as continual improvement. Here are a few questions that

ISO 9000 recommends which include a few of the expectations for nonconformance control from the nuclear industry:

3. Does the nonconformance reporting system prevent the installation or use of unsuitable products, components, materials, or services?

4. Are unacceptable activities, products or services identified? Documented and evaluated? Segregated if possible? Dispositioned by the responsible authority? Communicated to affected groups within the organization/the supplier/the customer?

5. Is there a formal process for root cause analysis of significant adverse conditions? Are there trained analysts?

A risk management program should be included in your approach to continual improvement. Under the Clean Air Act, the U.S. Environmental Protection Agency is directed to develop regulations and guidance for preventing, detecting and responding to the accidental release of certain substances from stationary sources (Shirley 1995). The Clean Air Act requires owners/operators of facilities to prepare and implement a risk management program if they have more than a threshold quantity of a regulated substance in a process.

There are three elements in a risk management program: (1) a hazard assessment to evaluate a worse-case accidental release; (2) an accidental release prevention plan; and (3) accidental release response procedures.

ENDNOTES

1. The standard for quality management systems, ISO 9000, recommends that policies be publicized. Rule of thumb in using this checklist: never stop at "YES/NO" answers. Identify the

how, why, when, where, etc. of observations and conclusions. For example, why is the documented policy adequate?

APPENDIX B

System Flowchart

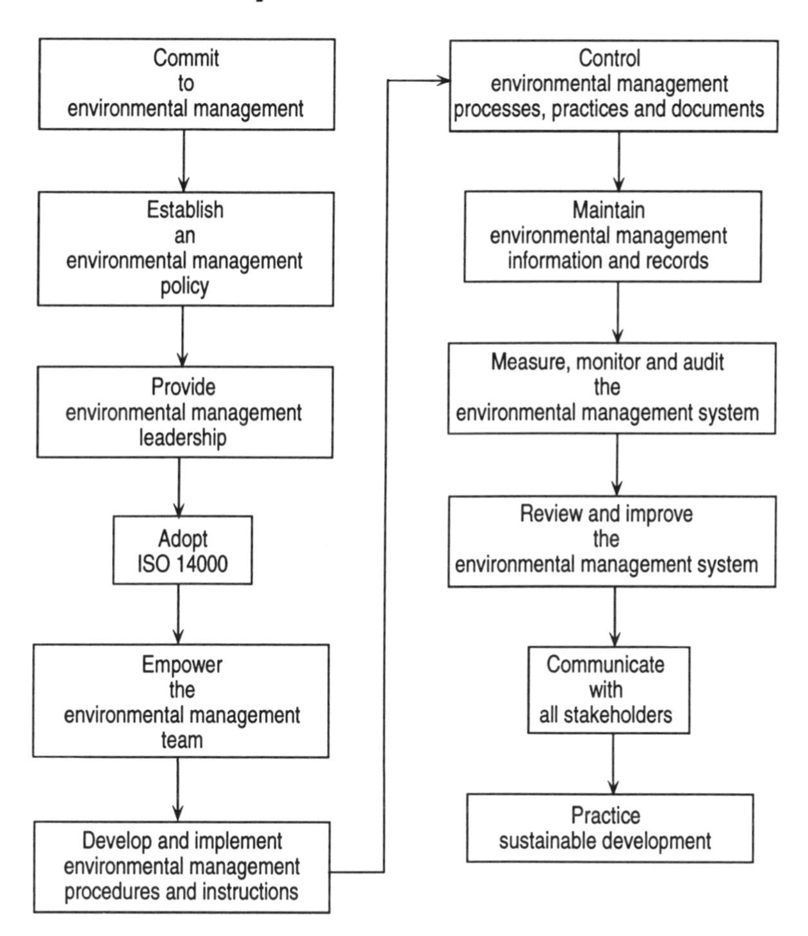
Commit
to
environmental management
Establish
an
environmental management
policy
Provide
environmental management
leadership
Adopt
ISO 14000
Empower
the
environmental management
team
Develop and implement
environmental management
procedures and instructions
Control
environmental management
processes, practices and documents
Maintain
environmental management
information and records
Measure, monitor and audit
the
environmental management system
Review and improve
the
environmental management system
Communicate
with
all stakeholders
Practice
sustainable development

Appendix C

Sample Procedure Format and Content

For years I have written, reviewed, audited, and walked down procedures in environmental protection, nuclear operation, corporate management, and emergency response. The following represents my personal preferences for effective procedures:

1. Start with the title. Make it as specific as you can regarding the function of the procedure but keep it to as few words as possible. For example: Design Control; Document Control; Control of Nonconformances; Audits; Surveillance; Records; etc.

2. Show the approval(s), the current revision, and the date effective on the cover sheet. Give each procedure a control number as unique identification.

3. State the purpose and scope of the procedure in the beginning. Be specific and be brief. For instance, "This procedure delineates the actions necessary to control the configuration of the facility. These actions are the responsibility of any individual causing a change to the as-built status of structures, systems, components or processes within the facility."

4. Identify key reference materials. For flowdown, what are the source documents (higher level procedures, policies, standards,

commitments, etc.)? Identify the use references, if any (other procedures required to complete the procedure, like the procedure for records, the procedure for controlling nonconformances, etc.).

5. Define key terms important to understand the procedure or make reference to a Glossary for specific terms.

6. Define key roles and responsibilities by position. Don't get elaborate and recreate the position descriptions, but tell the reader and the user of the procedure who does what.

7. State the required and expected action steps to accomplish the procedure. Short, simple, easy-to-comprehend action statements are encouraged. If you need to clarify a point or insert important points, use "NOTES," with double spacing before and after and maybe double underlining under the word "NOTES."

8. Identify the records generated by the procedure. Specify which are more important than others (like those records that provide evidence of compliance with the procedure and those that indicate use of the procedure).

9. Include necessary exhibits, enclosures, attachments, or appendices. Please be selective. Procedures should be as few pages as possible. Provide examples of forms to be used.

10. Flowchart each procedure but leave the flowcharts with the approval records OR mount the flowcharts near the work activity. Putting flowcharts in procedures seldom helps anyone but an auditor. Flowcharts should be simple and short. Flowcharts longer than a single page may indicate the procedure is too long, too complex, and likely to set the user up for failure.

Consider developing and using more instructions than procedures. There should be a single environmental management policy, a single vision statement, a single overall mission statement (there may be additional mission statements for specific components of your organization). Then there are sufficient procedures on a generic level, to flowdown requirements and commitments and to cover the basics. Finally, I suggest small, specific instructions (also controlled as documents) that fit the situation and the few users responsible. Had the nuclear industry used instructions more often, as the standards and regulations recommended and allowed, the cost of energy from the fission of atoms would be greatly reduced.

Procedures and instructions, like policy, should be written for the user. Use in-house talent if you can or get help from outside. Remember your audience. Involve the user from start to finish. Validate the defined process by flowchart and walkdown. Procedures and instructions are only as good as the end result they bring you.

Bibliography

Adair, John. 1990. *Understanding Motivation*. Guildford: Talbot Adair Press.

Allen, Roger E. 1994. *Winnie-the-Pooh on Management*. New York: Penguin Group.

ANSI/ASQC E4-1994, *Specifications and Guidelines for Quality Systems for Environmental Data Collection and Environmental Technology Programs.*

ANSI/ASME NQA-2, Part 2.7, *Software Quality Assurance.*

ANSI/ASQC Q9000-1-1994, *Quality Management and Quality Assurance Standard—Guidelines for Selection and Use.*

Arbuckle, J. Gordon and Thomas F.P. Sullivan. 1991. Fundamentals of Environmental Law. *Environmental Law Handbook*. Rockville, MD: Government Institutes.

ASME NQA-1-1989, *Quality Assurance Program Requirements for Nuclear Facilities.*

ASQC Energy and Environmental Division video, *Energy and Environmental Stewardship.*

ASQC *Quality Progress*. June 1995.

Bell, Chip R. 1994. *Customers as Partners: Building Relationships that Last a Lifetime*. San Francisco, CA: Berrett-Koehler Publishers.

Block, Peter. 1993. *Stewardship*. San Francisco, CA: Berrett-Koehler Publishers.

British Standard BS 7750:1994, *Specification for Environmental Management Systems*.

Burton, Terrence T. and John W. Moran. 1995. *The Future Focused Organization (Complete Organizational Alignment for Breakthrough Results)*. Englewood Cliffs, NJ: Prentice Hall.

Carr, Clay. 1992. *Smart Training, The Manger's Guide to Training for Improved Performance*. New York: McGraw-Hill.

Chemical and Engineering News. Vol. 72 (January 31, 1994): p. 17.

Cocheu, Ted. 1993. *Making Quality Happen (How Training Can Turn Strategy into Real Improvement)*. San Francisco, CA: Jossey-Bass Publishers.

Cohen, Allan R. 1993. *The Portable MBA in Management*. New York: J. Wiley & Sons.

Covey, Stephen R. 1989. *The Seven Habits of Highly Effective People, Restoring the Character Ethic*. New York: Simon & Schuster.

Currid, Cheryl and Co. 1994. *The Reengineering Toolkit, 15 Tools and Technologies for Reengineering Your Company*. Rocklin, CA: Prima Publishing.

DOE Order 5700.6C, *Quality Assurance*.

Ehrle, Carol. Full disclosure. *Resources*. June 1995: p. 13.

Environmental Action. Vol. 26 (Winter 1995): p. 6.

Export Today. May 1995: p. 5.

Fairley, Peter and Michael Roberts. Pilot projects and ISO 14000 moving forward but credibility remains elusive. *Chemical Week.* July 5/12, 1995: p. 19.

Fellers, Gary. 1992. *The Deming Vision: SPC/TQM for Administrators.* Milwaukee, WI: ASQC Quality Press.

Focus. Vol. 11 (6): p. 23.

Forsha, Harry. 1992. *The Pursuit of Quality through Personal Change.* Milwaukee, WI: ASQC Quality Press.

Gordon, Sean and Dan Tunstall. The Creation and Distribution of Environmental Information. *Environmental Information,* a Bulletin of the American Society for Information Science. April/May 1995: pp. 11–12.

Hadlet, Dave. *Interest in ISO 14000 standards. Quality Systems Update.* August 1995: p. 30.

Hawkin, Paul. The ecology of commerce. *Buzzworm, The Environmental Journal.* Vol. 16 (January 1994): p. 16.

Hitt, William D. 1990. *Ethics and Leadership, Putting Theory into Practice.* Columbus, OH: Batelle Press.

Industry Week. Vol. 244 (July 3, 1995): p. 9.

International Chamber of Commerce Business Charter for Sustainable Development.

ISO 14000: 199X, ISO/TC/207 SC1/WG2 N80 rev., Committee Draft prepared by ISO/TC207/SC1/WG2; February, 1995.

ISO 14001:199X, *Environmental Management Systems—Specification with Guidance on Use,* Committee Draft prepared by ISO/TC207/SC1/WG1; February, 1995.

ISO/TC/207 SC1/WG2 Committee Draft of February 1995.

Jick, Todd D. 1993. Managing Change. In *The Portable MBA in Management*, by Allan R. Cohen. New York: J. Wiley & Sons.

Kanter, Rosabeth Moss. 1993. Conclusion: Future Leaders Must Be Global Managers. In *The Portable MBA in Management*, by Allan R. Cohen. New York: J. Wiley & Sons.

Kennedy, Carol. 1991. *Instant Management: the Best Ideas from the People Who Have Made a Difference in How We Manage*. New York: W. Morrow & Co.

Kirschner, Elisabeth. *Chemical & Engineering News*. April 3, 1995: p. 13.

Luecke, Richard. 1994. *Scuttle Your Ships Before Advancing (and Other Lessons Learned from History on Leadership and Change for Today's Manager)*. New York: Oxford University Press.

McInerney, Francis and Sean White. 1993. *Beating Japan (How Hundreds of American Companies Are Beating Japan—and What Your Company Can Learn from Them)*. New York: Penguin Group.

Maniak, Angela J. 1990. *Report Writing for Internal Auditors*. Probus Publishing; Salem, Massachusetts.

Patterson, Susan M. and Danna A. Mancini. Fundamentally Fractal, Of Convoluted Curves and Transcendent Teams. In *The Future Focused Organization (Complete Organizational Alignment for Breakthrough Results)*, by T. Burton and J. Moran. Englewood Cliffs, NJ: Prentice Hall.

Perrone, Marissa A. and David Kirkpatrick. Green becomes standard. *Export Today*. May 1995: p. 13.

Power. Vol. 139 (8). August 1995: p. 10.

Quality Systems Update. August 1995: p. 15.

Richardson, Robert J. and S. Katherine Thayer. 1993. *The Charisma Factor (How to Develop Your Natural Leadership Ability)*. Englewood Cliffs, NJ: Prentice Hall.

Samdani, G. Sam, Stephen Moore, and Gerald Ondrey. ISO 14000: New passport to world markets. *Chemical Engineering*. June 1995.

San Diego Business Journal. 1994: pp. 14–15.

Shirley, William A. Regulatory update. *Chemical Engineering Progress*. July 1995: p. 13.

Sibson, Robert E. 1994. *Maximizing Employee Productivity, a Manager's Guide*. New York: Amacom Books.

Silverstein, Michael. 1993. *The Environmental Economic Revolution*. New York: St. Martin's Press.

Stewart, James R., Peter Mauch, and Frank Straka. 1994. *The 90-Day ISO Manual: The Basics*. Delray Beach, FL: St. Lucie Press.

Sutton, Gary, with Brian Tracy. 1993. *Tight Ships Don't Sink, Profit Secrets from a No-Nonsense CEO*. Englewood Cliffs, NJ: Prentice Hall.

Title 10 Code of Federal Regulations part 50, Appendix B, *Quality Assurance*.

Title 10 Code of Federal Regulations part 76, *Certification of Gaseous Diffusion Plants*.

Title 10 Code of Federal Regulations part 830, subpart 120, *Quality Assurance for Nuclear Facilities*.

United Nations. 1992. *Agenda 21*. New York: UNCED.

U.S. OMB Circular A-119. 1993. *Federal Participation in the Development and Use of Voluntary Standards*.

Vaill, Peter B. 1993. *Visionary Leadership*. In *The Portable MBA in Management*, by Allan R. Cohen. New York: J. Wiley & Sons.

Voehl, Frank, Peter Jackson, and David Ashton. 1994. *ISO 9000, An Implementation Guide for Small to Mid-Sized Businesses*. Delray Beach, FL: St. Lucie Press.

Additional Resources

Dale Carnegie books are a "must read." Nobody saw the executive's, the leader's, or the manager's role better. *How to Win Friends and Influence People* is a good start. It is a foundation of the Total Quality Management concepts currently in practice.

Important Organizations and Contacts

The International Organization for Standardization (ISO)
Rue de Varembé
CH-1211, Geneva 20
Switzerland
Phone: +41 22 749 01 11
Fax: +41 22 733 34 30

The ISO Central Secretariat
Case postale 56
CH-1211, Geneva 20
Switzerland
Phone: +41 22 749 01 11
Fax: +41 22 733 34 30

The U.S. Registration Accreditation Board, Inc. (RAB)
RAB
PO Box 3003
Milwaukee, Wisconsin 53201-3003
Phone: (800) 246-1948
Fax: (414) 272-1734

The British National Accreditation Council for Certification Bodies
19 Buckingham Gate
London SW1E 6LB
England
Phone: +44 (0) 71 233 7111
Fax: +44 (0) 71 233 5115

The Dutch Council for Certification (Raad Voor de Certifactie)
Stationsweg 13F,
3972 KA Driebergen
The Netherlands
Phone: +31 3438 12604
Fax: +31 3438 18554

The U.S. Department of Commerce European Community Affairs and International Trade Administration
14th Street & Constitution Avenue, NW
Washington DC 20230
Phone: (202) 482-2000

The British Standards Institution (BSI) offers overseas standards updating services and worldwide standards information.
Linford Wood
Milton Keynes MK14 6LE
England
Phone: +44 (0) 908 220022
Fax: +44 (0) 908 320856

The British Department of Trade and Industry (DTI)
Kingsgate House, Bay 511
66-74 Victoria Street
London SW1E 6SW
England
Phone: +44 (0) 71 215 8142
Fax: +44 (0) 71 215 8515

Standards New Zealand
NZS 9000-1
SNZ
Standards House
155 The Terrace
Wellington 6020 New Zealand
Phone: +64 4 498 5990
Fax: +64 4 498 5994

Earth Council
Apartado 2323
San José, Costa Rica
(For information on the national Councils for Sustainable Development in 132 countries)

INDEX

D

F

M

N

Q

R